Springer Series in Electrophysics
Volume 18   Edited by Walter Engl

Springer Series in Electrophysics

Editors: Günter Ecker   Walter Engl   Leopold B. Felsen

Takanori Okoshi

Planar Circuits
for Microwaves and Lightwaves

With 138 Figures

Springer-Verlag
Berlin Heidelberg New York Tokyo

Professor Dr. Takanori Okoshi

Department of Electronic Engineering, University of Tokyo
Bunkyo-ku, Tokyo 113, Japan

Series Editors:

Professor Dr. Günter Ecker

Ruhr-Universität Bochum, Theoretische Physik, Lehrstuhl I,
Universitätsstrasse 150, D-4630 Bochum-Querenburg, Fed. Rep. of Germany

Professor Dr. Walter Engl

Institut für Theoretische Elektrotechnik, Rhein.-Westf. Technische Hochschule,
Templergraben 55, D-5100 Aachen, Fed. Rep. of Germany

Professor Leopold B. Felsen Ph.D.

Polytechnic Institute of New York, 333 Jay Street, Brooklyn, NY 11201, USA

ISBN-13: 978-3-642-70085-9   e-ISBN-13: 978-3-642-70083-5
DOI: 10.1007/978-3-642-70083-5

Library of Congress Cataloging in Publication Data. Okoshi, Takanori, 1932-. Planar circuit for microwaves and lightwaves. (Springer series in electrophysics ; v. 18). Bibliography: p. Includes index. 1. Microwave integrated circuits. 2. Integrated optics. I. Title. II. Series. TK7876.0415 1985 621.381'32 84-22147

© Springer-Verlag Berlin Heidelberg 1985

Softcover reprint of the hardcover 1st edition 1985

Typesetting: K+V Fotosatz, 6124 Beerfelden.
Offset printing: Beltz Offsetdruck, 6944 Hemsbach/Bergstr.
Bookbinding: J. Schäffer OHG, 6718 Grünstadt.
2153/3130-543210

to my late mother

Preface

Until recently, three principal classes had been known in the electrical circuitry. They were as follows:

1) The *lumped-constant circuit*, which should be called a zero-dimensional circuit, in the sense that the circuit elements are much smaller in size as compared with the wavelength in all three spatial directions.
2) The *distributed-constant circuit*, which should be called a one-dimensional circuit, in the sense that the circuit elements are much smaller than the wavelength in two directions but comparable to the wavelength in one direction.
3) The *waveguide circuit*, which should be called a three-dimensional circuit, in the sense that the circuit elements are comparable to the wavelength in all three directions.

The principal subject of this book is the analysis and design (synthesis) theories for another circuit class which appeared in the late 1960s and became common in the 1970s. This new circuit class is

4) the *planar circuit*, which should be called a two-dimensional circuit, in the sense that the circuit elements are much smaller in size as compared with the wavelength in one direction, but comparable to the wavelength in the other two directions.

As described in the Introduction, the planar circuit concept has become not only technically significant, but also necessary in microwave engineering because of the advent of millimeter-wave integrated circuits and various low-impedance semiconductor microwave devices. The technical significance of the planar circuit is still increasing at present because it is now an inevitable tool in the *exact* analysis and design of microwave/millimeter-wave integrated circuits. Chapters 2 – 8 are devoted to the analysis and design of such "planar circuits" to be used in the microwave and millimeter-wave regions.

In the final two chapters (Chaps. 9, 10) a different circuit category is dealt with, that is,

5) the *optical planar circuit*, whose dimension is comparable to the wavelength in one derection, but is much larger than the wavelength in the other two directions.

Such a circuit category is also becoming technically significant because it plays an important role in some optical integrated circuits.

We should note that despite an apparent resemblance of their names, these two types of electromagnetic circuits, i.e., the planar circuit and the optical planar circuit, require entirely different analysis and design approaches.

The research of the planar circuit theory was started in Italy (S. Ridella et al.) and Japan (T. Okoshi) independently in 1968. In the author's case, the concept of the planar circuit first came to mind when he was asked to give a tutorial talk about the "general principle in the design of Gunn and IMPATT oscillator circuits" at a symposium held in a hilltop hotel in the Zaō Mountains, Japan, on August 28 – 29, 1968. At that time it was not yet easy to give a general guiding principle of the circuit design of this sort because the impedance characteristics of Gunn and IMPATT diodes had not been well known. The only undoubted fact about the impedance characteristics was that these devices had very low negative impedance as compared with electron-beam devices (tubes) which had been predominant in the microwave engineering until that time. Therefore, the author was obliged to construct the talk mainly on how to design stripline or waveguide circuits having very low impedance levels, for use in Gunn and IMPATT oscillators as their resonant circuits. Preparation for the talk led the author to the concepts of the open-boundary and short-boundary planar circuits having arbitrary shapes, which are described in Chaps. 3 and 4, respectively.

Throughout the 1970s and 1980s, many scientists in the world (Japan, Italy, the United States, Canada, India, Brazil, and so forth) joined the research of planar circuits. In addition to its analysis, even the computer-aided synthesis of planar circuits having prescribed characteristics has become possible, as will be found in Chaps. 6, 7. The system of theories of electrical circuitry, ranging from the zero-dimensional through three-dimensional, was completed finally by the addition of the planar circuit concept.

On the other hand, the significance of the concept of optical planar circuits, as defined above, came to be recognized by optoelectronics specialists in the 1970s. Frankly speaking, the author himself has contributed very little to the progress in this area. Descriptions in Chaps. 9, 10, therefore, are based principally upon the work performed by several excellent scientists in various countries of the world.

Tokyo, December 1984   *Takanori Okoshi*

Acknowledgments

The author is indebted to a large number of people for the work on which this book is based. In the earliest stage of the planar circuit research in my laboratory at the University of Tokyo, an important contribution was made by Professor Tanroku Miyoshi, presently with Kobe University, Japan. When I decided to start planar circuit research in 1968, he was a graduate student in my laboratory working for the M. S. degree. He got his M. S. degree in March 1969, and from April 1969 through March 1972 he concentrated entirely upon planar circuit research as his Ph. D. thesis work. His brilliance and effort brought forth various analysis techniques of open-boundary planar circuits described in Chaps. 2, 3.

Later, Professor Miyoshi and I wrote a book entitled *Planar Circuits* (in Japanese), which was published by Ohm Publishing Company, Tokyo, in 1975. In 1978, this book received the honor of being awarded the Excellent Book Prize by the Institute of Electronics and Communication Engineers of Japan (IECE Japan). Some parts of this English book are similar in content to the Japanese edition; these are the first halves of Chaps. 2, 3, and most of Chap. 4. (Other parts, i.e., about three-quarters of the entire volume, have been newly written on the basis of developments of the theory in the past nine years.) The present author would like to thank the coauthor of the previous Japanese edition, Professor Tanroku Miyoshi, who has generously permitted me to use in this book some figures and descriptions in the Japanese edition.

Since the early 1970s, a number of faculty members, assistants, graduate students, and undergraduate students joined the planar circuit research at the University of Tokyo. Only the names of the most important of these people are, in chronological order, Miss S. Kitazawa, Mr. T. Takeuchi, Mr. Y. Uehara, Prof. J.-P. Hsu, Prof. M. Saito, Dr. F. Kato, Mr. T. Imai, and Mr. K. Ito. In addition to the above people, Professors E. Yamashita of Electro-Communication University and Y. Kobayashi of Saitama University contributed much to the refinement of the theory through discussions at the "Planar Circuit Colloquium" held bimonthly for about ten years in Tokyo.

Finally, the author heartily thanks his secretaries, Miss M. Onozuka and Miss N. Matsunaga, who typed the manuscript and its versions many times. Their beautiful work encouraged the author greatly. Without their devoted assistance this book would never have been completed.

December, 1984   Takanori Okoshi

Contents

1.1.1 Conventional Four Ranks

We consider first four relatively conventional ranks. The circuitry which appeared first in the history of electrical engineering was the lumped-constant circuit, consisting of resistances, capacitances, and inductances. Next, in the middle of the nineteenth century, the distributed-constant line appeared: its theoretical treatment was investigated first in connection with the development of long-distance telegraph cables, in particular the first transatlantic submarine telegraph cable completed in 1858.

On the other hand, the theory of waveguide circuitry was developed almost 80 years later, in the 1930s and 1940s. This development owed much to the active microwave research during World War II aiming at the improvement of radars for military purposes.

After World War II, a significant trend in the electronic and communication engineering has been the extension of the practical frequency range into optical frequencies; in other words, the combination of electronics with optics. A new field called *optoelectronics* or *electrooptics* has been rapidly expanding. The conventional optical circuitry consisting of lenses, mirrors, and prisms has become familiar to electronics engineers as one class of electromagnetic circuitry. Such an electromagnetic circuit may be called a *free-space circuit*, in the sense that the size of the circuit is much greater than the wavelength in all three spatial directions.

1.1.2 Comparison of Circuit Dimensions

We compare here the above four kinds of electrical and optical circuitry with regard to their dimensions relative to the wavelength. Resonant circuits in the four circuit ranks are considered as examples.

In a lumped-constant resonator, the circuit elements (inductance L and capacitance C) are much smaller in size as compared with the wavelength λ in all three spatial directions x, y, and z. In a distributed-constant resonator, the circuit size is comparable to the wavelength (in most cases, approximately $\lambda/4$ or $\lambda/2$) in one direction, but much smaller in two directions normal to its length. A waveguide resonator has dimensions comparable to the wavelength in all three directions. In the optical free-space circuit, a typical resonator consists of two mirrors; this is called a Fabry-Perot resonator. Such a resonator has dimensions much greater than the wavelength in all three spatial directions.

The above four circuitries can be tabulated, as shown in Table 1.1 [1.1]. The lumped-constant circuit, in which $d_x, d_y, d_z \ll \lambda$, where d_x, d_y, and d_z denote sizes in the three directions, is considered to be the first rank. The waveguide circuit, in which $d_x, d_y, d_z \simeq \lambda$, is considered to be the fourth rank, whereas the optical free-space circuit, in which $d_x, d_y, d_z \gg \lambda$, is positioned at

1. Introduction

Three principal categories have been known in the electrical circuitry so far. They are the lumped-constant (zero-dimensional) circuit, distributed-constant (one-dimensional) circuit, and waveguide (three-dimensional) circuit. The planar circuit to be discussed in general in Chaps. 2 – 8 is a circuit category that should be positioned as a two-dimensional circuit. It is defined as an *electrical circuit having dimensions comparable to the wavelength in two directions but much less than it in the third direction.*

In addition to the planar circuit which is used mainly at microwave and millimeter-wave frequencies, the *optical planar circuit* is also a subject of this book and is dealt with in Chaps. 9, 10. It is defined as an *electromagnetic circuit having dimensions comparable to the wavelength in one direction, but much larger than it in the other two directions.*

In the present chapter, we first classify various electrical and electromagnetic circuits into seven ranks and investigate their features. In the latter half, technical features and the history of research of the planar circuit and the optical planar circuit are discussed in detail. Finally, the organization of this book is described briefly.

1.1 Seven Ranks in Electrical and Optical Circuitry

Electrical and optical circuits (electromagnetic circuits in general) can be classified, according to their circuit dimensions relative to the operating wavelength, into seven ranks:

1) lumped-constant circuits,
2) distributed-constant circuits,
3) planar circuits,
4) waveguide circuits,
5) long-waveguide circuits,
6) optical planar circuits,
7) optical free-space circuits.

Table 1.1. Classification (seven ranks) of electrical and optical circuits [1.1]

Rank	Classification	Dimensions with respect to the wavelength	Examples
1	Lumped-constant circuit	$d_x, d_y, d_z \ll \lambda$	Inductance, capacitance, and resistance
2	Distributed-constant circuit	$d_x, d_y \ll \lambda, d_z \simeq \lambda$	Coaxial resonator
3	Planar circuit	$d_z \ll \lambda, d_x, d_y \simeq \lambda$	Circular or square planar resonator
4	Waveguide circuit	$d_x, d_y, d_z \simeq \lambda$	Cavity resonator
5	Long-waveguide circuit	$d_x, d_y \simeq \lambda, d_z \gg \lambda$	The "sweeping comb filter" Monomode semiconductor laser
6	Optical planar circuit	$d_z \simeq \lambda, d_x, d_y \gg \lambda$	Some sort of optical planar circuit
7	Optical free-space circuit	$d_x, d_y, d_z \gg \lambda$	Ordinary optical systems using lenses and prisms

the seventh. According to such a ranking principle, the distributed-constant circuit, in which d_x, $d_y \ll \lambda$ but $d_z \simeq \lambda$, should be positioned as the second rank.

1.1.3 Nonconventional Three Ranks

Then what are the other three ranks, the third, fifth, and sixth ranks? These circuits had had relatively less technical significance until the 1970s. However, recent progress in electronics and optoelectronics has made these circuits technically more significant as seen in the following.

1.1.4 Planar Circuits

According to the above ranking principle, the third rank is defined as the circuit which has dimensions comparable to the wavelength in two directions but a much smaller thickness in one direction. Such circuits are now often found in microwave integrated circuits. This circuit class, the planar circuit, is the subject of Chaps. 2 – 8.

The planar circuit structure is often found in microwave integrated circuits (hereafter sometimes referred to as MICs), but more often in millimeter-wave integrated circuits (hereafter sometimes MMICs). The technical background follows. When the frequency increases, the width of striplines used in an MIC or MMIC cannot be reduced in proportion to the wavelength without an excessive increase of circuit loss. This statement will be explained in more detail. The starting question is: what should we do when the frequency is doubled in an MIC/MMIC design? For example, consider a branch-line 3-dB hybrid, as

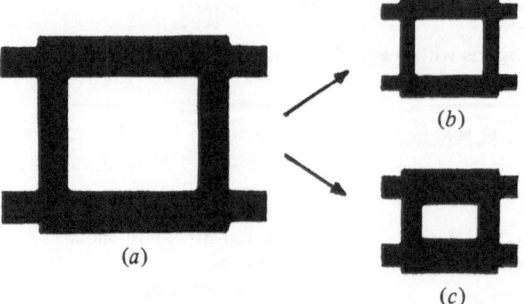

Fig. 1.1a–c. Possible modification of the pattern of a branch-line 3-dB hybrid when the frequency is doubled from f_0 (**a**) to $2f_0$ (**b**, **c**)

shown in Fig. 1.1, and assume that the impedance level should remain unchanged (e.g., 50 Ω) at the doubled frequency.

The first solution is to reduce every dimension of the circuit by half, in other words, to use a similar circuit pattern reduced by half with a half of the original substrate thickness (Fig. 1.1b). Thus the impedance level is kept unchanged. However, in such a case, the conductor loss per wavelength will increase in proportion to the square root of the frequency due to the skin effect.

Therefore, an alternative and practically better solution in most cases is to use a somewhat wider pattern with somewhat thicker substrate as compared with the strictly half-size circuit shown in Fig. 1.1b, the arm lengths being kept half-size. This solution is illustrated in Fig. 1.1c.

Thus, the stripline circuit pattern becomes wider (relative to the wavelength) as the frequency increases. Actually, in 3-dB couplers used in MMICs, the width of the stripline often becomes comparable to the arm wavelength ($\cong \lambda/4$) as seen in Fig. 1.2 [1.2]. Such a circuit can not be analyzed nor designed as a distributed-constant circuit because, in such a "widened" branch-line 3-dB coupler, we encounter a serious problem: how should we define the arm length?

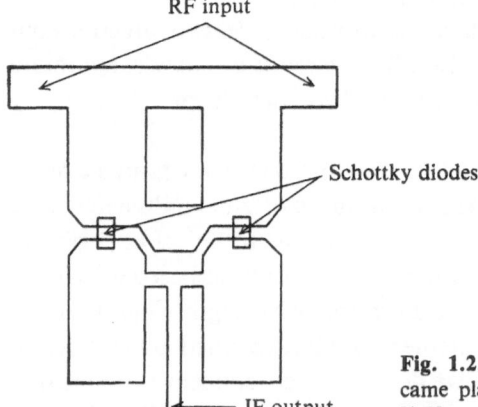

Fig. 1.2. An example of stripline circuits which became planar-like due to the impedance requirement [1.2]

Such a change in MIC/MMIC techniques was one of the two motivations that forced engineers to start planar circuit research in the late 1960s. The other motivation was the advent of various new semiconductor negative-impedance devices in microwave and millimeter-wave regions. Those devices such as Gunn, IMPATT, and TRAPATT diodes have impedance levels generally much lower than electron beams which offer negative impedance in electron tubes. Thus, circular, square, or triangular resonators became necessary as the external resonant circuits of microwave/millimeter-wave semiconductor oscillators so as to match the impedance level of the external circuit to that of the device. Such resonators also required analysis and design as planar circuits.

1.1.5 Optical Planar Circuits

The sixth category, which is characterized by its dimensions comparable to the wavelength in one direction but much greater than it in the other two directions, should be called the free-plane circuit, or more practically in the present state of the art, the optical planar circuit. This sort of circuit was first investigated by English investigators in the 1950s to construct a centimeter-wave spectrometer for measuring permittivity of dielectrics; they called it the *parallel plate spectrometer* because the electromagnetic energy was confined between two large (~1 square meter) copper plates [1.3]. However, this scheme has not become practical

This rank of circuit became common later at optical frequencies. The advent of the so-called integrated optics (or optical IC) in the 1970s stimulated the research and development of such circuits at optical frequencies; after the late 1970s these were sometimes called *optical planar circuits*. Such circuitry offered many interesting new technical problems. For example, we cannot use metallic boundaries effectively at optical frequencies. We have to rely more or less upon the relatively small difference of the refractive indices of two materials to confine the electromagnetic energy. This is what is actually done in optical planar circuits using glass substrates. The optical planar circuit is the subject of Chaps. 9, 10.

1.1.6 Long-Waveguide Circuit

The fifth category, the long-waveguide circuit (or long-waveguide resonator), in which d_x, $d_y \simeq \lambda$ but $d_z \gg \lambda$, is less important in the microwave area. The author and his colleagues have found that we can make an interesting use of it for measuring AM noise and FM noise, and the correlation between these, in the output of microwave oscillators [1.4]. Such a device was named the sweeping comb filter. At optical frequencies, resonators of this type are frequently used in monomode semiconductor lasers.

1.2 Classification and Technical Significance of Planar Circuits

1.2.1 Three Basic Structures

As described earlier, the planar circuit is defined as an *electromagnetic circuit having dimensions comparable to the wavelength in two directions but much less than it in the third direction.* The planar circuit thus defined has three possible basic configurations. They are the triplate type, open type, and short-boundary type, as shown in Fig. 1.3.

a) Triplate-Type Planar Circuit

An arbitrarily shaped center conductor plate is sandwiched by two ground conductors (Fig. 1.3a). The spacing material may be either air or some dielectrics. Actually, in the former case, the center conductor must be supported by thin dielectric blocks, or by a very thin dielectric plate as in "suspended-type" MICs.

A triplate-type planar circuit having a structure symmetrical with respect to the upper and lower ground conductors is of special theoretical importance. Most of the theoretical analyses described in the following chapters assume symmetrical modes in such a symmetrical structure.

b) Open-Type Planar Circuit

A structure in which one of the ground conductors in the triplate type is removed, or placed at a relatively distant place, is called open type (Fig. 1.3b). This corresponds to the so-called *microstrip structure* in the MIC design. Such a structure has disadvantages such as the loss increase due to the excitation of asymmetrical modes, the stray coupling between different parts of the circuit

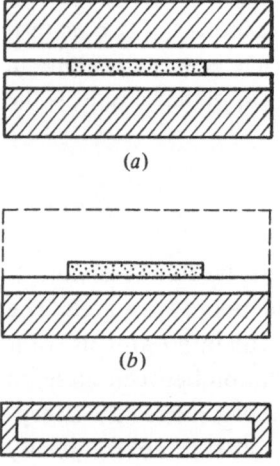

(a)

(b)

(c)

Fig. 1.3a – c. Classification of planar circuits: (a) triplate type, (b) open type (or unbalanced type), (c) short-boundary type

due to the spread of the electromagnetic field, and the difficulty in precise circuit analysis and design. Nevertheless, for practicality the planar circuit of this type is widely used because active devices can easily be built in, and *trimming* of the circuit is easy.

c) Short-Boundary-Type Planar Circuit

A very thin waveguide circuit having an arbitrary planar shape can also be considered to be a planar circuit. The planar circuit of this type has been investigated for a purpose entirely different from the above two types. The triplate type and open type have been investigated with their application to MICs in mind. In contrast to these, the theory of the waveguide-type planar circuit is useful in the analysis and design of ordinary microwave waveguide circuits based upon a TE_{10}-mode rectangular waveguide. This is because in such waveguide circuits the transverse electric field is often entirely absent; hence the planar circuit approach is mathematically valid even when the waveguide thickness is comparable to the wavelength (usually about $\lambda/3$).

In addition to the above three structures, a *biplate-type structure* is possible in which a dielectric plate is sandwiched by two arbitrarily shaped circuit conductors, like in a two-plate capacitor. Such a structure has been the subject of analysis by Italian investigators in the early stage of planar circuit research, with the name of *three-layer planar structure*. Such a structure is theoretically significant because it is the simplest planar circuit model. However, this has not been used in practical MICs. In the present book, it will be considered together with the triplate type because mathematical problems offered by these two types are entirely identical.

1.2.2 The Neumann and Dirichlet Problems

As described in the following chapters, the analysis of planar circuits is mathematically nothing but solving the two-dimensional wave equation (Helmholtz's equation) with appropriate boundary conditions. The triplate type has an open-circuited boundary; hence it offers the so-called *Neumann problem* in which the Helmholtz equation is solved with open (zero normal derivative) boundary conditions. On the other hand, the waveguide type has a short-circuited boundary; it offers a *Dirichlet problem* in which the Helmholtz equation is solved with short-circuited (zero amplitude) boundary conditions. The boundary conditions for the open type or for asymmetrical structures are more complex, but closer to the open-circuited rather than to the short-circuited one.

1.2.3 Technical Significance of the Planar Circuit Concept

The technical significance of the planar-circuit concept can be summarized in three items:

1) The planar circuit includes the stripline circuit as its special case. Therefore, in the MIC design the former offers a wider freedom than the latter. On the other hand, for highly advanced modern computers, the synthesis (design) as well as the analysis of an arbitrarily shaped planar circuit is easy. Therefore, the planar circuit concept now offers an exact and efficient tool in the MIC design.
2) The planar circuit can offer a lower impedance level than the stripline circuit does. Hence it matches microwave semiconductor devices such as Gunn and IMPATT diodes or the Schottky-barrier diode requiring low-impedance circuitry.
3) The system of theories of electrical circuitries, ranging from zero-dimensional through three-dimensional, can be completed by the addition of the planar circuit concept.

1.3 History of Planar Circuit Research

1.3.1 History of the Earliest Stage

The intensive research of planar circuits started late in the 1960s. However, we should note that at that time the planar circuit was already not an entirely new concept. A special case of this circuitry, the disk-shaped resonator, had been used in the stripline circulator or even as a filter [1.5, 6]. The so-called radial line, which is also a special case of the planar circuit, had been discussed as early as the 1940s [1.7].

However, until late in the 1960s, general treatment of the planar circuit, in other words, the analysis of arbitrarily shaped planar circuits, had never been presented. This sort of research started in Italy and Japan independently. In Italy, *Bianco, Biorci, Civalleri,* and *Ridella* showed the formulation of the impedance and admittance matrices of the biplate and triplate structure described in the last paragraph of Sect. 1.2.1, as well as its equivalent network [1.8 – 11]. In particular, they investigated intensively rectangular shaped triplate-type planar circuits and derived their impedance matrices as functions of frequency [1.11, 12].

In Japan, *Okoshi* pointed out the significance of the analysis and synthesis of two-dimensional circuits having arbitrary shapes, and named this circuit class the planar circuit [1.13 – 15]. Furthermore, *Okoshi, Migitaka* and *Miyazaki* used the planar circuit approach in the design of X-band Gunn oscillators [1.16]. *Okoshi* and *Miyoshi* presented an efficient method for the analysis of an arbitrarily shaped planar circuit, which is now called the contour-integral method, as well as the equivalent circuit parameters of rectangular and circular planar circuits [1.17 – 19].

The possibility and technical significance of the synthesis of planar circuits was also foreseen in the relatively early stage of the research [1.13, 14, 19]. In the conclusion of [1.19] it is stated, "What is emphasized is that we can analyze an arbitrarily shaped planar circuit within a reasonable computer time. The *design* of a planar circuit, based upon the high-speed computer analysis and the trial-and-error principle, will also be possible within several years". Actually, papers on the design (synthesis) started to appear in 1976 (Chaps. 6, 7).

1.3.2 History After 1972

After 1972, the number of researchers of the planar circuit increased steadily in Japan, Italy, Canada, India, the United States, Brasil, and many other countries. The basic framework of the theory of its analysis and synthesis was almost completed in the following ten years. The history in this stage, however, would become too lengthy if described in detail; it will be described in parts in the following chapters.

1.4 History of Optical Planar Circuit Research

The history of optical planar circuit research can be traced back to the late 1960s. Papers on optical ICs started to appear intensively in 1969. In 1970 and 1971, lightwave behavior in two-dimensional space such as a thin film or an optical IC was discussed [1.20]. The positioning of this circuitry with respect to other circuitries (such as shown in Table 1.1) was described first in 1969 [1.15].

1.5 Purpose and Organization of This Book

The major part of this book, Chaps. 2 – 8, deals with the planar circuit to be used in MICs or MMICs. The symmetrically excited open-boundary type is mainly considered because it is practically the most important and easiest to analyze among various types. The waveguide-type (short-boundary) planar circuit is also discussed in Chap. 4, but rather subsidiarily in the entire context of the description. Emphasis is placed upon the computer analysis of arbitrarily shaped planar circuits (Chaps. 3 – 5), and computer-aided synthesis of the planar circuit pattern giving prescribed circuit characteristics (Chaps. 6, 7). Chapter 8 presents the analysis and synthesis of anisotropic (ferrite) planar circuits.

The last two chapters (Chaps. 9, 10) are devoted to the analysis of optical planar circuits.

2. Analysis of Planar Circuits Having Simple Shapes

When the circuit pattern is relatively simple (for example, rectangular, circular, of annular), the equivalent circuit parameters of a triplate-type planar circuit can be obtained by an analytical approach based upon the Green's function of the wave equation. Such an analytical approach is described in this chapter to show the basic characteristics of the planar circuit.

The wave equation describing the wave behavior in such a circuit is derived. The solution of the wave equation is then obtained in the form of Green's function using an eigenfunction expansion. Circuit characteristics are derived from the Green's function of the circuit. Finally, examples of the circuit characteristics and some typical applications are described.

2.1 Background

The principal purpose of this chapter is to present the general formulation for the analysis of the network characteristics of a symmetrically excited, open-boundary (i.e., triplate-type) planar circuit, as shown in Fig. 2.1a. The term "analysis" denotes here the determination of the circuit parameters of the equivalent multiport, as shown in Fig. 2.1b. The actual circuit characteristics and some applications of triplate planar circuits having simple shapes (rectangular, circular, triangular, and annular) will be treated, too.

The basic equation describing the electromagnetic field in the planar circuit is a two-dimensional wave equation (Helmholtz equation). The circuit characteristics are obtained as the solution of the Helmholtz equation under given boundary conditions. As described in Sect. 1.2.2, in a triplate-type planar circuit, the boundary condition upon the periphery of the circuit is given as an open-circuited one except at coupling ports.

The solution of a Helmholtz equation under open-boundary conditions can be derived without difficulty if the Green's function of the second kind (i.e., the Neumann type) for the Helmholtz equation is given. The discussion in this chapter is limited to planar circuits having simple shapes, for which the Green's functions can easily be derived.

To understand fully the behavior of a planar circuit, we should know the electromagnetic field distribution inside the circuit pattern as well as along its

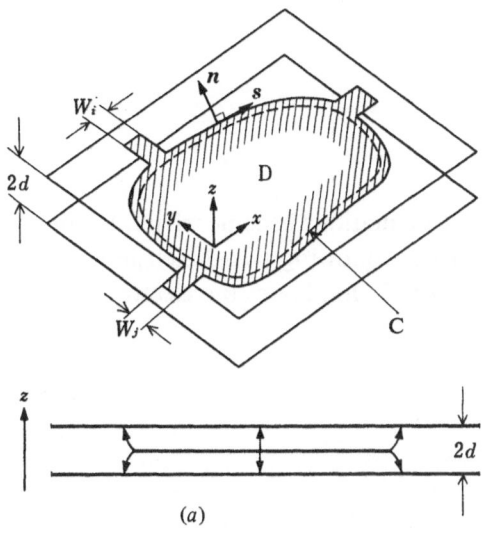

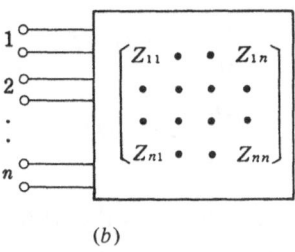

(b)

(a)

Fig. 2.1a, b. A triplate-type planar circuit: (a) physical structure, (b) the equivalent multiport

periphery. However, for practical purposes, the knowledge about the electromagnetic field distribution inside the circuit is not very important, because the equivalent circuit parameters (Fig. 2.1b) are determined by the field distribution along the periphery. The following analysis [2.1] is performed with the above fact in mind.

2.2 Basic Equations

2.2.1 Wave Equation

A symmetrically excited, triplate planar circuit, as shown in Fig. 2.1a, is considered throughout this chapter.

An arbitrarily shaped, thin conductor circuit plate is sandwiched by two ground conductors, with a spacing d from each of them. The circuit is assumed to be excited symmetrically with respect to the upper and lower ground conductors. There are several coupling ports, and their widths are denoted by W_i, W_j, ... The rest of the periphery is assumed to be open-circuited. The x and y coordinates and the z axis, respectively, represent a set parallel and perpendicular to the conductors.

When the spacing d is much smaller than the wavelength and the spacing material is homogeneous and isotropic, we can assume that $\partial/\partial z = 0$ and $H_z = E_x = E_y = 0$. Therefore, Maxwell's equations are simplified as

$$\frac{\partial H_y}{\partial x} - \frac{\partial H_x}{\partial y} = j\omega\varepsilon E_z, \tag{2.1}$$

$$\frac{\partial E_z}{\partial y} = -j\omega\mu H_x,\tag{2.2}$$

$$\frac{\partial E_z}{\partial x} = j\omega\mu H_y.\tag{2.3}$$

Note that the other three of the Maxwell's equations become trivial. From the above three equations, it is found that the following two-dimensional wave equation (Helmholtz equation) in terms of E_z dominates the electromagnetic field in the planar circuit:

$$(\nabla_T^2 + k^2)E_z = 0,\quad\text{where}\tag{2.4a}$$

$$k = \omega\sqrt{\varepsilon\mu} = \omega/c = 2\pi/\lambda,\tag{2.4b}$$

$$\nabla_T^2 = \frac{\partial^2}{\partial x^2} + \frac{\partial^2}{\partial y^2},\tag{2.4c}$$

and ω, ε, μ, and k denote the angular frequency, permittivity and permeability of the spacing material, and the wavenumber in it, respectively.

In most of the discussion in this book, the circuit is assumed to be lossless. However, when we consider a small circuit dissipation, the wavenumber k should be replaced formally by

$$k = k' - jk'',\quad (k' \gg k'')\tag{2.5a}$$

where

$$k' = \omega\sqrt{\varepsilon\mu},\tag{2.5b}$$

$$k'' = \omega\sqrt{\varepsilon\mu}\,(\tan\delta + r/d)/2.\tag{2.5c}$$

Here δ is the loss angle of the spacing material [$\delta = \tan^{-1}(\sigma/\omega\varepsilon)$], and r is the skin depth of the conductor. The derivation of (2.5) is shown in Appendix A2.1. In the following, however, we assume that the circuit is lossless and k is real.

2.2.2 Boundary Conditions for Cases when External Ports Are Absent

Next we consider the boundary conditions. The first necessary step is to investigate the behavior of the surface current on the circuit plate. If we denote the surface current density by $i(i_x, i_y)$,

$$H_y = i_x\quad\text{and}\quad H_x = -i_y\tag{2.6}$$

hold for a radio-frequency field. Therefore, we obtain from (2.1) and (2.6)

$$\text{div}\,\boldsymbol{i} = \mathrm{j}\,\omega\varepsilon E_z\,. \tag{2.7}$$

To begin with, we consider a case where not external circuit is connected to the circuit periphery, and hence no current flows into (or flows out from) the circuit along its periphery. On the other hand, (2.6) shows that everywhere in the circuit, the magnetic field $\boldsymbol{H}$ and surface current density $\boldsymbol{i}$ are orthogonal with each other. Therefore, along the open-circuited periphery the magnetic field lines intersect the periphery normally; in other words, the open boundary constitutes the so-called "magnetic wall".

On the other hand, (2.2, 3) tell us that the direction of the gradient of E_z is normal to $\boldsymbol{H}$. Therefore, the boundary condition along the open-boundary periphery of the circuit is given as

$$\frac{\partial E_z}{\partial n} = \frac{\partial V}{\partial n} = 0\,, \tag{2.8}$$

where n denotes the local coordinate outward normal to the periphery, and V denotes the rf (radio-frequency) voltage of the center conductor with respect to the ground conductors, i.e., $V = E_z d$. Thus the problem to be solved is formulated as

$$(\nabla_T^2 + k^2)\,V = 0 \quad (\text{in } D)\,, \tag{2.9}$$

$$\frac{\partial V}{\partial n} = 0 \qquad (\text{on } C)\,, \tag{2.10}$$

where C and D denote the periphery and the region inside the periphery, respectively.

2.2.3 Eigenfunction Expansion

Equations (2.9, 10) describe an eigenvalue problem. In other words, the non-trivial solution (nonzero solution) of (2.9) is present only for an infinite number of discrete values of k. These values k_0, k_1, k_2, ... are called eigenvalues; they correspond to resonant angular frequencies ω_0, ω_1, ω_2, ... where $\omega_i = ck_i$ (c: the velocity of light). To each eigenvalue k_i, a specific field solution ϕ_i exists; this is called the eigenfunction.

Incidentally, as shown in Appendix A2.2, eigenvalues k_i's are given as stationary values of a variational expression

$$k^2(V) = \frac{\iint\limits_D (\nabla V)^2 dS}{\iint\limits_D V^2 dS}\,, \quad (\nabla V = \text{grad } V) \tag{2.11}$$

whereas the eigenfunctions ϕ_i's satisfy the following orthogonal relations:

$$\iint\limits_{D} \phi_n \phi_m dS = \delta_{nm} , \qquad (2.12)$$

where δ_{nm} is the Kronecker's delta, i.e., $\delta_{nm} = 1$ for $n = m$ and $\delta_{nm} = 0$ for $n \neq m$. In the above equation, ϕ_i's are assumed to be normalized. Any existing solution V can be expressed by a linear combination of a set of ϕ_i's.

The above relations (2.11, 12) will be discussed further and used later in Chap. 4. In this chapter, we will try first to give a comprehensive picture of the problem rather than to be involved in mathematical details.

2.2.4 A Simple Example of the Solution

As the simplest example of the planar circuit, we consider first a rectangular circuit plate having dimensions $a \times b$ and no external ports. Such a circuit plate and the two ground conductors constitute a resonator. If we define the (x, y) coordinates as shown in Fig. 2.2, the stationary (nontransmitting) solution of the wave equation (2.9) is given as

$$\phi(x, y) = V_0 \begin{Bmatrix} \sin \\ \cos \end{Bmatrix} k_x x \begin{Bmatrix} \sin \\ \cos \end{Bmatrix} k_y y , \qquad \text{where} \qquad (2.13a)$$

$$k_x^2 + k_y^2 - k^2 = 0 . \qquad (2.13b)$$

The values of k_x and k_y are given as

$$k_x = m\pi/a , \qquad (2.14a)$$

$$k_y = n\pi/b , \qquad (2.14b)$$

where m and n denote integers, because of the boundary conditions given as $\partial\phi/\partial x = 0$ at $x = 0$ and $x = a$, and $\partial\phi/\partial y = 0$ at $y = 0$ and $y = b$. For the same reason only cosine in (2.13a) applies. Hence, we obtain

$$\phi(x, y) = \frac{2}{\sqrt{ab}} \cos \frac{m\pi x}{a} \cos \frac{n\pi y}{b} , \qquad (2.15)$$

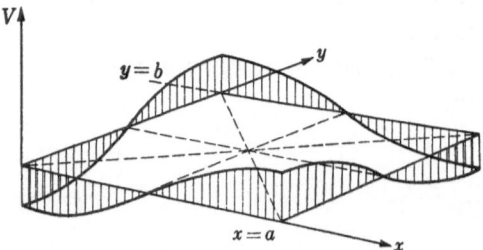

Fig. 2.2. An example of resonance modes in a rectangular resonator (quadrupole mode)

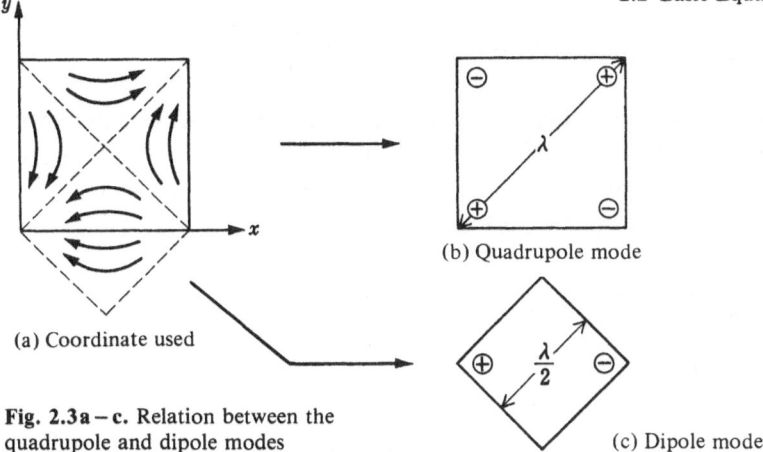

(a) Coordinate used

(b) Quadrupole mode

(c) Dipole mode

Fig. 2.3a–c. Relation between the
quadrupole and dipole modes

where the coefficient is chosen so as to satisfy the ortho-normalizing condition
(2.12). The resonance takes place only at wavelengths λ which satisfy

$$\left(\frac{m}{a}\right)^2 + \left(\frac{n}{b}\right)^2 = \left(\frac{k}{\pi}\right)^2 = \left(\frac{2}{\lambda}\right)^2 .\tag{2.16}$$

As the simplest example, we consider a square circuit ($a = b$) and the case
$m = n = 1$. Then we have

$$a = b = \sqrt{2}\,\pi/k = \lambda/\sqrt{2} .\tag{2.17}$$

The above equation tells that this ($m = 1$, $n = 1$) mode of resonance takes
place at the wavelength equal to the diagonal of the square circuit pattern. It is
also found in (2.15) that the resonance mode in this case is the so-called
quadrupole mode, as shown in Fig. 2.2.

We should note that the (1, 1) mode is not the fundamental mode resonance. This fact is easily understood if we sketch the surface current distribution on the circuit, as shown in Fig. 2.3a. This figure shows that we can cut
out a smaller square having size $(\lambda/2) \times (\lambda/2)$, on which a "dipole mode" can
be excited at the same frequency. This means that for a square circuit of a
given size, such a dipole mode gives the lowest-frequency resonance.

The dipole mode (Fig. 2.3c) has been derived in a spatially *tilted* direction.
It is funny that this mode can never be expressed in an *upright* direction by a
single mode function given by (2.15). However, it can be expressed in an upright direction as the superposition of two modes, i.e., ($m = 1$, $n = 0$) and
($m = 0$, $n = 1$) modes, as

$$\phi(x, y) = \frac{2}{u} \cdot \left(\cos\frac{\pi x}{a} + \cos\frac{\pi y}{a}\right) .\tag{2.18}$$

In this case the length of the side along x and y axes is equal to $\lambda/2$.

2.2.5 Boundary Condition at Ports

We consider next the case where external ports are present. At junctions with external circuits, (2.8) is no longer valid. Equations (2.2 or 3 and 6) suggest that, if an external circuit having admittance Y is connected to this port, Y is related to the voltage V at the port as

$$Y = \frac{2\int_W i_n ds}{\int_W V ds/W} = \frac{2W\int_W H_s ds}{\int_W V ds} = \frac{-2j W\int_W \left(\frac{\partial V}{\partial n}\right) ds}{\omega\mu d\int_W V ds} , \qquad (2.19)$$

where W denotes the width of the port, i_n is the outward surface current density across the contour C, and H_s is the magnetic field intensity along C. In (2.19), the factor 2 expresses the fact that the current flows on both the upper and lower surfaces of the circuit plate in the triplate structure. Thus, the external admittance is related to the averages of V and $(\partial V/\partial n)$ over the port width W.

2.3 Derivation of Circuit Characteristics

2.3.1 Definition of Terminal Voltage and Current

Next we consider a planar circuit having several ports, as shown in Fig. 2.4. We assume here that the widths of the ports are all much smaller than the wavelength, and define the terminal voltage V and terminal current I as averages over the port widths:

$$V = d\int_W E_z ds/W, \quad I = 2\int_W i_n ds . \qquad (2.20)$$

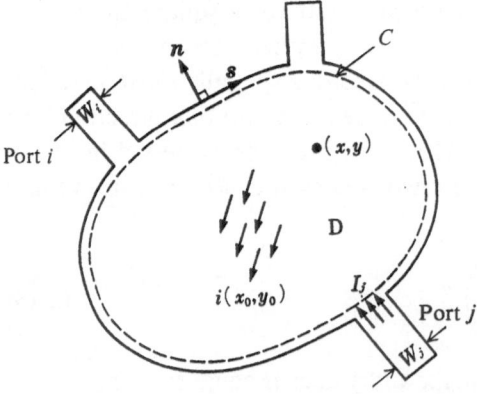

Fig. 2.4. Symbols used in the definitions of terminal voltage and current, and Green's function analysis

In the following we show briefly that the terminal voltage and current, as defined above, satisfy the reciprocity conditions. We consider first a case when current flows only into the pth port while other ports are open circuited, and denote the electric and magnetic fields by $E^{(1)}$ and $H^{(1)}$. Similarly, we denote the fields by $E^{(2)}$ and $H^{(2)}$ for the case when current flows only into the qth port. From Maxwell's equations, it can be shown without difficulty that when the dielectric spacing material is isotropic,

$$\nabla \cdot (E^{(1)} \times H^{(2)} - E^{(2)} \times H^{(1)}) = 0 \tag{2.21}$$

holds [2.2]. Integrating (2.21) and using Gauss's theorem, we obtain

$$\oint_C (E^{(1)} \times H^{(2)} - E^{(2)} \times H^{(1)}) \cdot n \, ds = 0 . \tag{2.22}$$

However, the line integral is required only over the widths of ports; hence

$$\int_{W_q} i_{n_q}^{(2)} E_z^{(1)} ds = \int_{W_p} i_{n_p}^{(1)} E_z^{(2)} ds . \tag{2.23}$$

This equation shows the reciprocal nature of the field.

When the widths of ports are much smaller than the wavelength, the electric fields $E_z^{(1)}$ at port q and $E_z^{(2)}$ at point p can be assumed to be constant over the widths W_q and W_p, respectively. Hence, we can substitute (2.20) into (2.23), to obtain with good accuracy

$$I_q^{(2)} V_q^{(1)} = I_p^{(1)} V_p^{(2)} . \tag{2.24}$$

This relation can be rewritten, using relations $Z_{pq} = V_q^{(1)}/I_p^{(1)}$ and $Z_{qp} = V_p^{(2)}/I_q^{(2)}$, as

$$(Z_{qp} - Z_{pq}) I_p^{(1)} I_q^{(2)} = 0 . \tag{2.25}$$

Therefore, because $I_p^{(1)}$ and $I_q^{(2)}$ are arbitrary,

$$Z_{pq} = Z_{qp} . \tag{2.26}$$

Thus, it is proved that the terminal voltage and current defined as (2.20) make the corresponding impedance matrix $\{Z_{pq}\}$ symmetrical (i.e., $Z_{pq} = Z_{qp}$) with good accuracy.

2.3.2 Circuit Characteristics Expressed in Terms of Green's Function

Next, we derive the expression for Z_{pq} as defined above in terms of the circuit pattern. For this purpose we introduce Green's function G of the second kind, having a dimension of impedance which satisfies

$$V(x, y) = \iint_D \mathcal{G}(x, y \,|\, x_0, y_0)\, i(x_0, y_0)\, dx_0\, dy_0 \tag{2.27}$$

inside the periphery C shown in Fig. 2.4, and an open boundary condition

$$\partial \mathcal{G}/\partial n = 0 \tag{2.28}$$

along C. The term "second kind" is customarily used to denote an open-boundary condition. In (2.27), $i(x_0, y_0)$ denotes an assumed (fictitious) rf current density injected normally into the circuit (Fig. 2.4).

In ordinary triplate-type planar circuits, however, current is injected not normally, but from the periphery where coupling ports are present. Moreover, as to the rf voltage V, we are usually concerned only with the voltage along the periphery. In such a case, we can express the voltage along the periphery by a line integral in terms of the current density along the periphery as

$$V(s) = -\oint_C \mathcal{G}(s \,|\, s_0)\, i_n(s_0)\, ds_0, \tag{2.29}$$

where s and s_0 are used to denote distance along C, and i_n is the line current density outward normal to C at coupling ports.

Equation (2.29) shows that the rf voltage along the periphery is determined if the current density along the periphery is given, provided that the Green's function is known. Since i_n is present only at coupling ports, the integral in (2.29) should be performed only over the widths of the ports. Thus, we can compute the terminal voltage defined as (2.20), i.e., $V_i = \int_{W_i} V(s)\, ds/W$, directly from (2.29) as

$$V_i = \frac{1}{W_i} \sum_j \int_{W_i} \int_{W_j} \mathcal{G}(s \,|\, s_0)\, i_n(s_0)\, ds_0\, ds$$

$$\cong \sum_j \frac{I_j}{2 W_i W_j} \int_{W_i} \int_{W_j} \mathcal{G}(s \,|\, s_0)\, ds_0\, ds, \tag{2.30}$$

where $I_j = -2 \int_{W_j} i_n(s_0)\, ds_0$ represents the current flowing into the jth port on both the upper and lower surfaces of the circuit plate. In deriving the approximate form in (2.30), we assume that W_i, $W_j \ll \lambda$, and hence $\mathcal{G}$ does not vary appreciably within W_i and W_j, so that the average value of $\mathcal{G} i_n$ can be approximated as the product of the average of $\mathcal{G}$ and the average of i_n.

Equation (2.30) shows that the (i, j) element of the impedance matrix of a N-port planar circuit, i.e., $Z_{ij} = V_i/I_j$, is given as

$$Z_{ij} = \frac{1}{2 W_i W_j} \int_{W_i} \int_{W_j} \mathcal{G}(s \,|\, s_0)\, ds_0\, ds. \tag{2.31}$$

This is the basic equation of the triplate-type planar circuit.

When the shape of the circuit is relatively simple (rectangular of circular, for example), we can obtain the Green's function of the wave equation analytically, and derive the equivalent circuit parameters directly from that Green's function using (2.31). In the following subsection we investigate how to derive the Green's function.

2.3.3 Expansion of Green's Function by Eigenfunctions

Let the eigenvalues and eigenfunctions of the wave equation (2.9) be denoted by k_n and ϕ_n, respectively. As stated in Sect. 2.2.3, any function satisfying (2.9) can be expanded by a set of ϕ_n. Therefore, the Green's function $\mathcal{G}(x, y \,|\, x_0, y_0)$ can be expressed as [2.3]

$$\mathcal{G}(x, y \,|\, x_0, y_0) = \sum_n A_n \phi_n . \tag{2.32}$$

The remaining problem is how to compute coefficients A_n.

As a preparatory step, we show first that $\mathcal{G}$ satisfies a differential equation

$$(\nabla^2 + k^2) \, \mathcal{G} = -j\omega\mu d\delta(x - x_0)\,\delta(y - y_0) , \tag{2.33}$$

where δ denotes a delta function. To prove the above equation, we integrate this after multiplying both sides by the fictitious injection current $i(x_0, y_0)$. Then we obtain from the left-hand side

$$\iint (\nabla^2 + k^2) \, \mathcal{G}(x, y \,|\, x_0, y_0)\, i(x_0, y_0)\, dx_0 dy_0 = (\nabla^2 + k^2) V(x, y) ,$$

and from the right-hand side

$$-j\omega\mu d \iint i(x_0, y_0)\, \delta(x - x_0)\, \delta(y - y_0)\, dx_0 dy_0 = -j\omega\mu d i(x, y) .$$

Therefore, (2.33) can be proved if we can show that, if the fictitious injection current $i(x_0, y_0)$ were present, $(\nabla^2 + k^2) V = 0$ would become

$$(\nabla^2 + k^2) V = -j\omega\mu d i(x, y) . \tag{2.34}$$

This equation can readily be proved if we rewrite (2.1), adding $-i(x, y)$ to the displacement current density $j\omega\varepsilon E_z$, as

$$\frac{\partial H_y}{\partial x} - \frac{\partial H_x}{\partial y} = j\omega\varepsilon E_z - i , \tag{2.35}$$

and modify the computation performed to derive (2.4a) from (2.1). Thus (2.33) has been proved.

Next, we substitute (2.32) into (2.33) and use $(\nabla^2 + k_n^2)\phi_n = 0$ to obtain

$$\sum_n A_n(k^2 - k_n^2)\phi_n = -j\omega\mu d\delta(x - x_0)\delta(y - y_0) . \tag{2.36}$$

We multiply both sides of this equation by ϕ_m, integrate those in the region D, and finally use the ortho-normalizing condition given by (2.12). Then, the Green's function is given in terms of the eigenfunctions as

$$\mathscr{G}(x, y \,|\, x_0, y_0) = j\omega\mu d \sum_n \frac{\phi_n(x_0, y_0)\,\phi_n(x, y)}{k_n^2 - k^2} . \tag{2.37}$$

Setting (2.37) into (2.31), we find that the element of the impedance matrix z_{ij} is expressed, in a partial fraction form in terms of k, as

$$Z_{ij} = \frac{1}{2W_i W_j} \int_{W_i} \int_{W_j} j\omega\mu d \sum_n \frac{\phi_n(s_i)\,\phi_n(s_j)}{k_n^2 - k^2} \, ds_i ds_j . \tag{2.38}$$

This equation shows that when k is close to one of the k_n's, the corresponding term predominates because $(k_n^2 - k^2)$ in the denominator becomes very small. Therefore, when we are concerned only with the circuit characteristics at frequencies close to a resonance, we can pick up a predominating term from (2.38) to simplify the analysis. Generally, however, we have to perform a more complicated calculation considering at least several modes; this usually requires a computer even when the circuit pattern is simple.

2.4 Examples of Analysis Based on Green's Function

In this section, analyses of the simplest (rectangular, circular, triangular, and annular) circuits will be presented. Examples of more complicated circuit patterns will be shown in Chap. 3.

2.4.1 Rectangular Circuit

We consider a rectangular circuit having dimensions $a \times b$. The ortho-normalized eigenfunctions satisfying the boundary conditions are given as

$$\phi_{mn}(x, y) = \frac{\varepsilon_m \varepsilon_n}{\sqrt{ab}} \cos(k_{xm}x)\cos(k_{yn}y) ,$$

$$k_{xm} = m\pi/a, \qquad\qquad k_{yn} = n\pi/b \tag{2.39}$$

$$\varepsilon_m = \begin{cases} 1 & (\text{for } m = 0) \\ \sqrt{2} & (\text{for } m \neq 0) \end{cases} \qquad \varepsilon_n = \begin{cases} 1 & (\text{for } n = 0) \\ \sqrt{2} & (\text{for } n \neq 0), \end{cases}$$

where ε_m and ε_n are coefficients to make ϕ_{mn} satisfy the normalizing condition (2.12). The above equation leads directly to a Green's function

$$\mathscr{G}(x, y \,|\, x_0, y_0) = \frac{j\omega\mu d}{ab} \sum_{m=0}^{\infty} \sum_{n=0}^{\infty} \varepsilon_m^2 \varepsilon_n^2 \frac{\cos(k_{xm}x_0)\cos(k_{yn}y_0)}{k_{xm}^2 + k_{yn}^2 - k^2}$$

$$\times \cos(k_{xm}x)\cos(k_{yn}y) . \tag{2.40}$$

As an example, let us compute the input impedance Z_{in} of one-port rectangular circuit as shown in Fig. 2.5a. We obtain, using (2.31, 40),

$$Z_{\text{in}} = \sum_{m=0}^{\infty} \sum_{n=0}^{\infty} \frac{j\omega\mu d\varepsilon_m^2 \varepsilon_n^2}{2ab(k_{xm}^2 + k_{yn}^2 - k^2)} (\cos k_{xm} T)^2 \left(\frac{\sin k_{xm} W/2}{k_{xm} W/2} \right)^2 . \tag{2.41a}$$

If we set (2.5a) for k into (2.41a) to take the circuit dissipation into account, we can rewrite (2.41a), after some algebraic computations, as

$$Z_{\text{in}} = \sum_{m=1}^{\infty} \sum_{n=1}^{\infty} \frac{1}{j\omega C_{mn} - j(1/\omega L_{mn}) + G_{mn}} + \frac{1}{j\omega C_{00} + G_{00}} . \tag{2.41b}$$

In this expression, the second term $(j\omega C_{00} + G_{00})^{-1}$ stems from the mode with $m = n = 0$. This mode corresponds to the charging and discharging of the static capacitance of the circuit, and may be called the zero-frequency resonance or electrostatic mode. Parameters C_{00} and G_{00} express the static capacitance and the associated dielectric loss, and are given as

$$C_{00} = 2\varepsilon ab/d , \tag{2.42a}$$

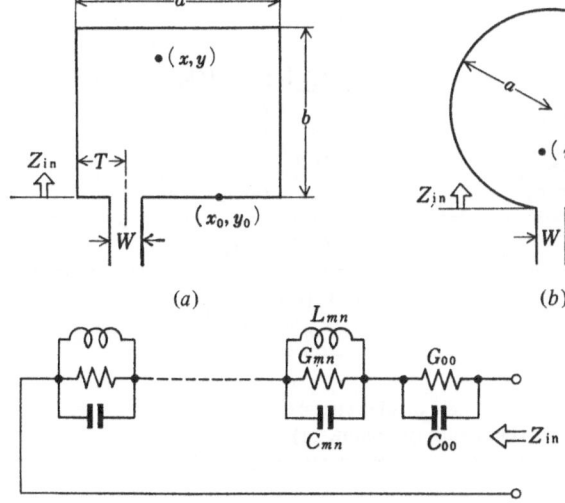

(a)

(b)

(c)

Fig. 2.5a–c. One-port planar resonators and their equivalent circuit: (a) one-port rectangular resonator, (b) one-port circular resonator, (c) equivalent circuit of one-port planar resonator

$$G_{00} = C_{00}(\tan \delta + r/d) , \tag{2.42b}$$

where r denotes the skin depth of the conductors used in the circuit. Equation (2.42b) can be derived without difficulty from (2.5a – c), by using $(\omega C_{00} + G_{00}) = \omega C_{00}(k/k')^2$ and assuming that $k' \gg k''$.

Equation (2.41b) shows that the equivalent circuit representing the input impedance is given, as shown in Fig. 2.5c, by a series connection of a number of parallel resonant circuits corresponding to each mode including the zero-frequency resonance ($m = n = 0$). Parameters C_{mn}, L_{mn}, and G_{mn} in (2.41b) give directly the equivalent circuit parameters in each resonant circuit. These parameters can be computed in terms of geometrical parameters by using (2.5a) and (2.41b), and are tabulated in the left half of Table 2.1, except for the electrostatic mode parameters, which are shown in (2.42a, b).

The coefficient F in Table 2.1 correspond to the final factor in (2.41a), which appears as the result of the integration of (2.31) over the port width W. Obviously this expresses the effect of finiteness of W (i.e., $W \neq 0$), and approaches unity, as W tends to zero. In three equations giving Q_0 in Table 2.1, Q_d and Q_c denote the dielectric-loss and conductor-loss Q factors, respectively. These relations would be obvious if we note (2.42b) and that $Q_0^{-1} = G_{00}/\omega C_{00}$.

Table 2.1. Equivalent circuit parameters of a rectangular circuit (Fig. 2.5a) and a circular circuit (Fig. 2.5b)

	Rectangular circuit	Circular circuit
Eigenfunct.	$\cos(k_{xm}x)\cos(k_{yn}y)$	$J_m(k_{mn}r)\cos m\theta$
Res. freq. f_{mn}	$\dfrac{\sqrt{(m/a)^2 + (n/b)^2}}{2\sqrt{\varepsilon\mu}}$	$\dfrac{k_{mn}}{2\pi\sqrt{\varepsilon\mu}}$
C_{mn}	$\dfrac{2\varepsilon ab}{\varepsilon_m 2\varepsilon_n 2d} \cdot \dfrac{1}{\cos^2(k_{xm}T)} \cdot \dfrac{1}{F}$	$\dfrac{2\pi a^2 \varepsilon[1 - m^2/(ak_{mn})^2]}{\varepsilon_m 2d} \cdot \dfrac{1}{F}$
L_{mn}	$\dfrac{\varepsilon_m 2\varepsilon_n 2\mu d \cos^2(k_{xm}T)}{2ab[(m\pi/a)^2 + (n\pi/b)^2]} F$	$\dfrac{\varepsilon_m 2\mu d}{2\pi} \cdot \dfrac{1}{(ak_{mn})^2 - m^2} F$
G_{mn}	$2\pi f_{mn} C_{mn}/Q_0$	$2\pi f_{mn} C_{mn}/Q_0$
F	$\left(\dfrac{\sin(k_{xm}W/2)}{k_{xm}W/2}\right)^2$	$\left(\dfrac{\sin(mW/2a)}{mW/2a}\right)^2$
Q_0	$Q_0^{-1} = Q_d^{-1} + Q_c^{-1}$ $Q_d = 1/\tan\delta$ (δ: loss angle of dielectrics) $Q_c = d/r$ (r: skin depth of conductor)	

2.4.2 Circular Circuit

Next we consider a circular circuit with radius a, as shown in Fig. 2.5b. The ortho-normalized eigenfunctions are given as

$$\phi_{mn}(r, \theta) = \frac{\varepsilon_m J_m(k_{mn}r) \cos m(\theta - \theta_0)}{a\sqrt{\pi(1 - m^2/a^2 k_{mn}^2)} J_m(k_{mn}a)} \begin{pmatrix} m = 0, 1, \ldots \\ n = 1, 2, \ldots \end{pmatrix} \tag{2.43a}$$

$$(\varepsilon_m = 1 \quad \text{for} \quad m = 0, \quad \varepsilon_m = \sqrt{2} \quad \text{for} \quad m \neq 0),$$

$$\phi_{00}(r, \theta) = \frac{1}{\sqrt{\pi a^2}} \quad (m = n = 0), \tag{2.43b}$$

where J_m denotes the mth order Bessel function of the first kind, and k_m is the nth root of $[(\partial/\partial r) J_m(kr)]_{r=a} = 0$. Equation (2.43b) again expresses the electrostatic mode corresponding to the charging and discharging of the static capacitance. The coefficient ε_m is again introduced so that ϕ_{mn} satisfies the normalizing conditions (2.12). Note that those modes for which $n = 0$ and $m = 1, 2, 3, \ldots$ never exist.

The Green's function can be obtained, by setting (2.43a, b) into (2.37), as

$$\mathscr{G}(r, \theta \mid r_0, \theta_0) = \sum_{m=0}^{\infty} \sum_{n=1}^{\infty} \frac{\varepsilon_m^2 j \omega \mu d J_m(k_{mn}r_0) J_m(k_{mn}r) \cos m(\theta - \theta_0)}{\pi(k_{mn}^2 - k^2) a^2 \left(1 - \dfrac{m^2}{a^2 k_{mn}^2}\right) J_m^2(k_{mn}a)}$$

$$+ \frac{d}{j\varepsilon\pi a^2}. \tag{2.44}$$

The equivalent circuit parameters of a one-port circular circuit can be computed by using (2.31, 44) in the same way as in the case of the rectangular circuit. The results obtained are tabulated in the right half of Table 2.1, again except for the electrostatic mode $m = n = 0$. The values of $k_{mn}a$ for each mode, given as the nth root of $[(\partial/\partial r) J_m(kr)]_{r=a} = 0$, are tabulated in Table 2.2. Parameters C_{00} and G_{00} are given as $C_{00} = 2\varepsilon\pi a^2/d$ and $G_{00} = C_{00}(\tan\delta + r/d)$, respectively, where ε denotes the permittivity of the spacing material.

Table 2.2. The nth root of $J_m'(x) = 0$

n \ m	0	1	2	3	4	5	6	7
1	3.832	1.841	3.054	4.201	5.317	6.416	7.501	8.578
2	7.016	5.331	6.706	8.015	9.282	10.520	11.735	12.932
3	10.173	8.536	9.969	11.346	12.682	13.987		
4	13.324	11.706	13.170					

2.4.3 Triangular Circuit

Triangular circuits are not frequently used in microwave/millimeter-wave ICs. *Helszajn* and *James* described a possibility of constituting bandpass or bandstop filters (Fig. 2.6) [2.4]. However, such circuits have not commonly been used. Moreover, the analysis of the triangular circuit is rather tedious except for some special cases such as $60° - 60° - 60°$ (equilateral), $30° - 60° - 90°$ (half equilateral), and $45° - 45° - 90°$ (isosceles right-angled) triangles [2.5]. Therefore, in this section we restrict ourselves to the equilateral triangle and derive its eigenvalues and eigenfunctions. The Green's function and the impedance matrix can easily be obtained from these if the circuit size and positions of ports are given.

The characteristics of the triangular planar circuit were investigated in general probably first by Y. Mizoguchi in 1972 in his graduation thesis at the University of Tokyo; the results he obtained were briefly introduced in [2.1]. Later, *Helszajn* and *James* [2.4], and *Chadha* and *Gupta* [2.5] investigated this circuit in more detail. In this section the description follows mainly [2.1].

We consider an equilateral triangle having side length a and the radius of the inscribed circle b (Fig. 2.7). We denote the three apexes by A, B, and C, the origin by O, and angles made by the x axis and AO, BO, CO by α, β, γ, respectively, as shown in Fig. 2.7. We next introduce the following triangular coordinates:

$$u = x \cos \alpha + y \sin \alpha$$
$$v = x \cos \beta + y \sin \beta \quad (\beta = \alpha + \tfrac{2}{3}\pi) \tag{2.45}$$
$$w = x \cos \gamma + y \sin \gamma \quad (\gamma = \beta + \tfrac{2}{3}\pi) .$$

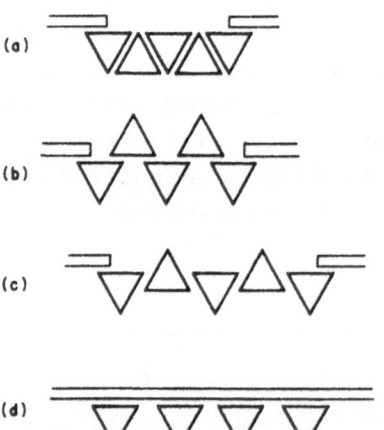

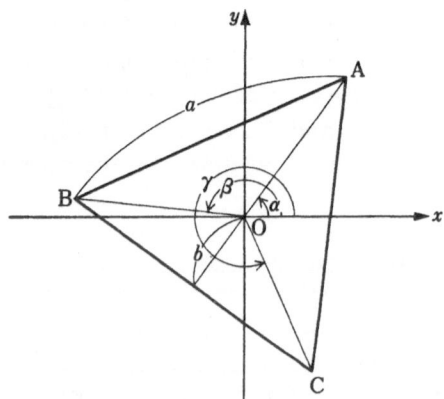

Fig. 2.6a – d. Possible configurations of bandpass filters (a – c), and bandstop filter (d) using triangular resonators [2.4]

Fig. 2.7. Symbols used in the analysis of an equilateral triangular circuit

If we define three variables u, v, w as above, the contour $u = $ const is normal to AO, and hence parallel to BC. Similarly, contours $v = $ const and $w = $ const are parallel to CA and AB, respectively.

Suppose we draw three lines, all passing a point (x, y), and parallel to each side of the triangle. Then u, v, and w denote the distances from the origin O to these lines, respectively. Hence, the three sides are expressed by $u = b$, $v = b$, and $w = b$, respectively; the open-boundary conditions are expressed as $\partial V/\partial n = 0$ for $u = b$, $v = b$, and $w = b$.

If we solve the Helmholtz equation (2.9) under such conditions, we obtain the eigenfunctions ϕ_{mn} as [2.6]

$$
\begin{aligned}
\phi_{mn}(x, y) = &\cos\left[\frac{2\pi l}{3b}\left(\frac{u}{2}+b\right)\right]\cos\left(\frac{\pi(m-n)(v-w)}{9b}\right) \\
&+\cos\left[\frac{2\pi m}{3b}\left(\frac{u}{2}+b\right)\right]\cos\left(\frac{\pi(n-l)(v-w)}{9b}\right) \\
&+\cos\left[\frac{2\pi n}{3b}\left(\frac{u}{2}+b\right)\right]\cos\left(\frac{\pi(l-m)(v-w)}{9b}\right)
\end{aligned}
\tag{2.46a}
$$

for cases except $m = n = l = 0$, whereas

$$
\phi_{00} = (4/\sqrt{3}a)^{1/2},
\tag{2.46b}
$$

and the corresponding eigenvalues as

$$
k = \frac{2\pi}{3\sqrt{3}b}\sqrt{m^2+mn+n^2} = \frac{4\pi}{3a}\sqrt{m^2+mn+n^2}.
\tag{2.47}
$$

In the above equations, parameters l, m, n are integers which satisfy $l+m+n = 0$. The (0, 0, 0) mode (2.46b) is again the electrostatic mode. Note that (2.46a) is not normalized because of complexity.

Six degenerated fundamental modes exist, which are given by

$$
\begin{aligned}
m &= 1, & n &= 0, & l &= -1 \\
m &= 0, & n &= 1, & l &= -1 \\
m &= 1, & n &= -1, & l &= 0 \\
m &= 0, & n &= -1, & l &= 1 \\
m &= -1, & n &= 1, & l &= 0 \\
m &= -1, & n &= 1, & l &= 1\,.
\end{aligned}
\tag{2.48}
$$

These are all "modified dipole modes" as shown in Fig. 2.8a, whose resonance frequency and wavelength are given as

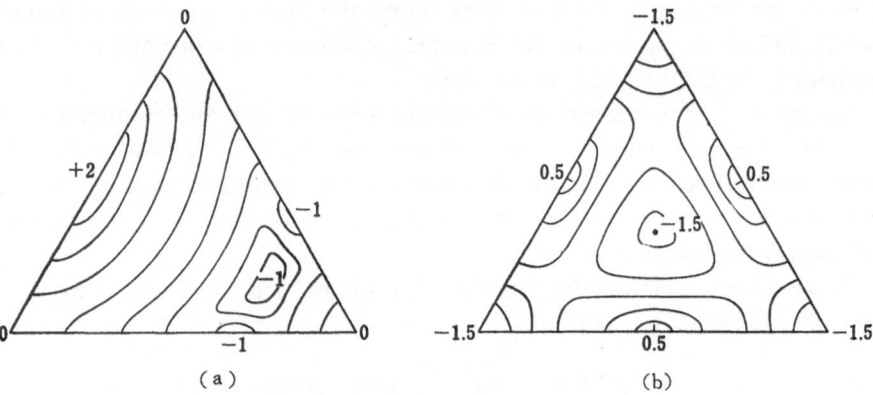

Fig. 2.8. The rf voltage distribution in a equilateral triangular circuit: (a) modified dipole mode, (b) tripole mode

$$k = 4\pi/3a, \quad \text{and} \tag{2.49a}$$

$$\lambda = 3a/2, \tag{2.49b}$$

respectively.

The second-lowest modes are given by the following six sets of l, m, and n:

$$
\begin{aligned}
m &= 1, & n &= 1, & l &= -2 \\
m &= -1, & n &= -1, & l &= 2 \\
m &= 1, & n &= -2, & l &= 1 \\
m &= -2, & n &= 1, & l &= 1 \\
m &= -1, & n &= 2, & l &= -1 \\
m &= 2, & n &= -1, & l &= -1 .
\end{aligned}
\tag{2.50}
$$

These modes are all "tripole modes" (Fig. 2.8b), whose resonance frequency and wavelength are given as

$$k = 4\pi/\sqrt{3}a, \quad \text{and} \tag{2.51a}$$

$$\lambda = \sqrt{3}a/2, \tag{2.51b}$$

respectively.

Another sort of triangular circuit that permits relatively simple analysis is the isosceles triangle. This has already been considered briefly in Sect. 2.2.4 (Fig. 2.3a).

2.4.4 Annular Circuit

We consider an annular circuit having inner and outer radii, a and b, respectively, as shown in Fig. 2.9. The ortho-normalized eigenfunctions of such a circuit are given as [2.1]

$$\phi_{mn}(r, \theta) = \frac{1}{K_{mn}} \left[J_m(k_{mn}r) - \frac{J_m(k_{mn}a)}{Y_m(k_{mn}a)} Y_m(k_{mn}r) \right] \cos \left[m(\theta - \theta_0) \right]$$

(2.52a)

$$(m = 0, 1, 2, \ldots , \quad n = 1, 2, 3, \ldots)$$

$$\phi_{00}(r, \theta) = 1/K_{00},$$

(2.52b)

where the second eigenfunction (2.52b) again corresponds to the electrostatic mode. Similar to the case of the circular circuit (Sect. 2.4.2), those modes for which $n = 0$ and $m = 1, 2, 3, \ldots$ never exist. Parameters k_{mn} in (2.52a) denote the nth roots of

$$J'_m(kb) Y_m(ka) - Y'_m(kb) J_m(ka) = 0.$$

(2.53)

Here Y_m is the mth order Bessel function of the second kind, and a prime denotes a differentiation.

Constants K_{mn} and K_{00} are introduced in Eqs. (2.52a, b) to normalize the eigenfunctions. From the normalizing condition

$$\int_0^{2\pi} d\theta \int_a^b \phi_{mn}^2(r, \theta) r \, dr = 1,$$

(2.54)

we obtain

$$K_{mn} = \varepsilon_m \frac{\pi k_{mn}^2}{4} \left(\frac{1}{Y_m^2(k_{mn}b)} - \frac{1}{Y_m^2(k_{mn}a)} \right)^{-1} \quad \begin{pmatrix} m = 0, 1, 2, \ldots \\ n = 1, 2, \ldots \end{pmatrix},$$

where

(2.55a)

$$\varepsilon_m = \begin{cases} 1 & (\text{for } m = 0) \\ \sqrt{2} & (\text{for } m \neq 0), \quad \text{and} \end{cases}$$

(2.55b)

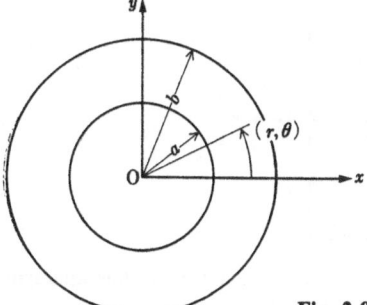

Fig. 2.9. An annular circuit and symbols used in analysis

$$K_{00} = 1/\sqrt{\pi(b^2 - a^2)} \ . \tag{2.55c}$$

The annular circuit itself is scarcely used in actual MICs (microwave integrated circuits). Instead, annular sectors as well as circular sectors are frequently found in actual MIC circuit components.

2.4.5 Circular and Annular Sectors

Circular and annular sectors were investigated by *Chadha* and *Gupta* [2.7]. Figure 2.10 shows some of the practical cases in which such planar circuit elements constitute microstrip bends and tapers. In [2.7] Green's functions for a circular sector and an annular sector are derived. However, the details are omitted here because of space limitations.

2.4.6 Open-Ended Stripline
(Comparison with Distributed-Constant Line Theory)

Finally, an open-ended stripline (Fig. 2.11) is analyzed to compare the result with that of distributed-constant line theory.

The input impedance of an open-ended stripline having width a and length b is given, by substituting $W_i = W_j = a$ into (2.31), as

$$Z_{\text{in}} = \frac{1}{2a^2} \int_0^a \int_0^a \mathscr{G}(s \,|\, s_0) \, ds_0 \, ds \ . \tag{2.56}$$

On the other hand, the Green's function can readily be obtained by letting $y = y_0 = 0$ and $m = 0$ in (2.40). Substituting this Green's function into (2.56), we obtain

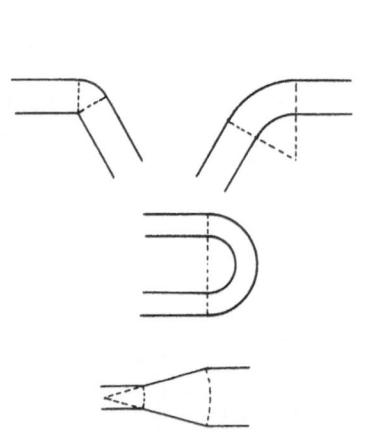

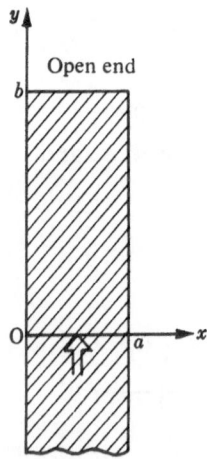

Fig. 2.10. Use of circular and annular sectors in stripline bends and taper section [2.7]

Fig. 2.11. An open-ended stripline and symbols used in analysis

$$Z_{in} = \frac{1}{2a^2} \int_0^a \int_0^a \frac{j\omega\mu d}{ab} \sum_{n=0}^{\infty} \frac{\varepsilon_n^2}{k_y^2 - k^2} dx\, dx_0$$

$$= \frac{j\omega\mu d}{2ab} \left(\frac{1}{-k^2} + \sum_{n=1}^{\infty} \frac{2}{k_y^2 - k^2} \right) = -j Z_0 \cot kb \,, \tag{2.57}$$

where $Z_0 = (d/2a)\sqrt{\mu/\varepsilon}$, and this gives the characteristic impedance of a tri-plate planar line having width a.

The final form in (2.57) shows that the input impedance derived from the planar circuit theory agrees with that given by the distributed-constant line theory.

2.5 Determination of Equivalent Circuit Parameters Based on Energy Consideration

The equivalent circuit of a planar resonator shown in Fig. 2.5c suggests that when the resonance frequencies are widely separated and the Q factor of the resonance is relatively high, the circuit characteristics in the vicinity of a res-onant frequency can be expressed approximately in terms of a set of L, C, and G. So far we have learned how to compute the resonant frequency f_0 and the unloaded Q factor, i.e., Q_0. We cannot determine three circuit parameters L, C, and G from two parameters, f_0 and G_0. In the following it is shown how these three parameters can be determined from f_0, Q_0, and the energy stored in the circuit.

We compute first the stored energy W_T for the quadrupole mode ($m = n = 1$) in a triplate-type square circuit as an example. We assume that the coupling port is positioned at one of the four corners. Generally, in a resonant circuit, the average electric energy $\langle W_E \rangle$, average magnetic energy $\langle W_M \rangle$, peak electric energy W_{Ep}, and the total energy W_T are related as

$$W_T = \langle W_E \rangle + \langle W_M \rangle = 2 \langle W_E \rangle, \tag{2.58}$$

$$\langle W_E \rangle = \tfrac{1}{2} W_{Ep} = \tfrac{1}{2} W_T \,. \tag{2.59}$$

Therefore, in the following we compute W_{Ep} instead of W_T.

The subscripts p and rms are used to denote peak and root-mean-square values. The circuit voltage at the corner of the square is denoted as V_{corner}. First, we may write W_{Ep} as

$$W_{Ep} = \tfrac{1}{2}\varepsilon d \int\int E_{zp}^2 dS \times 2 \,, \tag{2.60}$$

where E_{zp} denotes the peak electric field in the z direction. The factor 2 in (2.60) stems from the fact that we have upper and lower spaces in a triplate planar circuit. We note next that for any eigenmode satisfying $m \geqslant 1$ and $n \geqslant 1$,

$$\iint E_{zp}^2 ds = \frac{1}{4} S \left(\frac{V_{corner,p}}{d} \right)^2 \tag{2.61}$$

holds, where S denotes the circuit area. Therefore,

$$W_T = W_{Ep} = \frac{\varepsilon S}{4d} V_{corner,p}^2 = \frac{\varepsilon S}{2d} V_{corner,rms}^2 . \tag{2.62a}$$

On the other hand, the capacitance C in the equivalent circuit should be related to W_T as

$$W_T = 2 \langle W_E \rangle = C V_{corner,rms}^2 . \tag{2.62b}$$

Hence, combining (2.62a, b) and (2.42a), we obtain

$$C = \varepsilon S/2d = C_{00}/4 . \tag{2.63}$$

The factor of (1/4) originates from the same factor in (2.61); this is nothing but $(1/\varepsilon_m^2 \varepsilon_n^2)$.

Once the parameter C is thus determined, and if we suppose that f_0 and Q_0 are already known, we may compute other circuit parameters L and G by using the following relations:

$$2\pi f_0 \sqrt{LC} = 1, \quad \text{and} \tag{2.64}$$

$$Q_0 = 2\pi f_0 C/G . \tag{2.65}$$

It is easily found that the values of C, L, and G thus determined agree with those shown in Table 2.1 except for the factor F, which is a correction factor for the finite width of the coupling port. As seen in Table 2.1, F is given as a squared sampling function in terms of W, and hence approaches unity when W becomes infinitesimally small.

A numerical example is presented in the following. We consider a triplate-type square circuit having $a = 1$ cm, d (spacer thickness) $= 0.5$ mm, $\tan \delta = 10^{-4}$, and $\varepsilon_s = 2.25$, and compute its equivalent circuit parameters C, L, and G for the dipole-mode resonance, assuming that the coupling port is located at the corner. The computed parameters are as follow:

λ_0 (resonant frequency) $= 2a\sqrt{\varepsilon_s} = 3$ cm $\therefore$ $f_0 = 10$ GHz

$C_{00} = 2\varepsilon_0 \varepsilon_s S/d = 15.8$ pF $\therefore$ $C = 3.95$ pF

$L = 1/\omega^2 C \cong 0.064$ nH

$Q_d = 10\,000, \quad Q_c = d/r \cong 1000$ $\therefore$ $Q_0 = 910$

$G = \omega C/Q_0 = 0.27$ mS .

2.6 Equivalent Circuit of a Multiport Planar Circuit

We go back now to (2.38) and derive the equivalent circuit for a multiport planar circuit. If we separate the electrostatic mode from other rf modes, we may write an impedance matrix element as

$$Z_{ij} = \frac{1}{2W_iW_j} \int_{W_i} \int_{W_j} j\omega\mu d \sum_n \frac{\phi_n(s_i)\phi_n(s_j)}{k_n^2 - k^2} ds_i ds_j - \frac{j\omega\mu d}{2(\omega/c)^2} S. \qquad (2.66)$$

Therefore, when the circuit can be assumed to be lossless, we may write Z_{ij} as

$$Z_{ij} = \sum_{n=1}^{\infty} \frac{N_{ni}N_{nj}}{j\omega C_n + 1/j\omega L_n} + \frac{1}{j\omega C_0}, \qquad \text{where} \qquad (2.67)$$

$$C_n = 2\varepsilon S/d, \qquad L_n = d/2\omega_n^2 \varepsilon S, \qquad C_0 = 2\varepsilon S/d, \qquad (2.68)$$

$$N_{ni} = \frac{\sqrt{S}}{W_i} \int_{W_i} \phi_n(s_i) ds_i, \qquad N_{nj} = \frac{\sqrt{S}}{W_j} \int_{W_j} \phi_n(s_j) ds_j. \qquad (2.69)$$

When the circuit is not lossless, (2.67) should be corrected, to take into account the effect of loss, as

$$Z_{ij} = \sum_{n=1}^{\infty} \frac{N_{ni}N_{nj}}{j\omega C_n + 1/j\omega L_n + G_n} + \frac{1}{j\omega C_0 + G_0}. \qquad (2.70)$$

Here G_n is expressed for each of the resonant modes as

$$G_n = \omega_n C_n/Q_{0n}, \qquad (2.71)$$

where Q_{0n} denotes the corresponding unloaded Q factor, and

$$G_0 = C_0 \tan\delta. \qquad (2.72)$$

From the above calculations, the lumped-constant equivalent circuit of a multiport planar circuit can be expressed as Fig. 2.12, where N_{ni} and N_{nj} de-

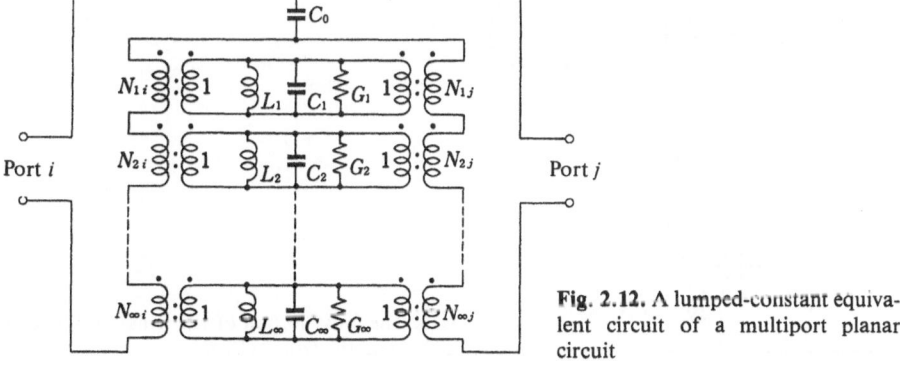

Fig. 2.12. A lumped-constant equivalent circuit of a multiport planar circuit

note step-up (or step-down) ratios of ideal transformers. Figure 2.12 shows a two-port circuit; however, this can easily be modified to show an N-port circuit by connecting the third, fourth, and further external circuits in parallel across the resonant circuits.

2.7 Validity of the Open-Boundary Assumption

So far we have assumed that at those places along the circuit periphery where no coupling port exists, the boundary condition for the wave equation is given as a magnetic wall, in other words, as a complete open boundary ($\partial V/\partial n = 0$). Actually some correction is necessary to take into account the so-called fringe effect.

To deal with the fringe effect in a quantitative manner, we tentatively consider a circuit fringe which is straight and uniform in x direction, as shown in Fig. 2.13. We easily see in this figure that the circuit boundary itself does not strictly constitute a magnetic wall, but the electromagnetic fields are extended toward the outside region. When both the triplate structure and the circuit excitations are symmetrical with respect to the circuit plate, such an extended field never excites a propagating mode, but decays rapidly outward.

The configuration of the decaying fringe field was analyzed in [2.8]. The important results follow. The magnetic field along the circuit periphery H_x satisfies $(\partial^2/\partial y^2 + \partial^2/\partial z^2 + k^2)H_x = 0$. We may solve this equation using Schwarz-Christoffel transform if we neglect the thickness of the circuit plate. We may further calculate the susceptance jB seen outwards (in the positive z direction) at Plane A in Fig. 2.13. The result is, for unit length along the periphery,

$$jB \cong j\sqrt{\frac{\varepsilon}{\mu}}\tan\left[\frac{2kd}{\pi}\ln 2 + S_1\left(\frac{2kd}{\pi};0,0\right) - 2S_1\left(\frac{kd}{\pi};0,0\right)\right], \quad (2.73)$$

where $k = \omega\sqrt{\varepsilon\mu} = 2\pi/\lambda$, and S_1 denotes a hypergeometric function defined as

$$S_1(x;0,0) = \sum_{n=1}^{\infty}\left(\sin^{-1}\frac{x}{n} - \frac{x}{n}\right). \quad (2.74)$$

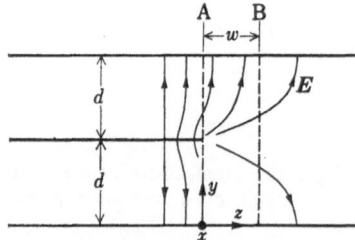

Fig. 2.13. The cross section of the fringe of a planar circuit showing the fringe effect

Here we consider how to express the above additional susceptance by an extention of the periphery from Plane A to B shown in Fig. 2.13. The position of Plane B, which is denoted by w in the figure, is given, from (2.74), as

$$w = \frac{1}{\pi}\left[2d\ln 2 + \lambda S_1\left(\frac{4d}{\lambda}; 0, 0\right) - 2\lambda S_1\left(\frac{2d}{\lambda}; 0, 0\right)\right]. \qquad (2.75)$$

When $4d \ll \lambda$, i.e., the conductor spacing is much smaller than the wavelength, as is mostly the case, the second and third terms of (2.75) can be neglected, so that

$$w = 2d\ln 2/\pi = 0.442 d. \qquad (2.76)$$

Note that the above w is irrelevant to the wavelength, and agrees with the value computed from the increase of the static capacitance due to the fringe effect [2.9].

2.8 Examples of Planar Circuits Having Simple Shapes

Three examples of planar circuits having simple shapes will be described; their design theories and experimental data will be presented and compared.

2.8.1 Circular Resonator

We first consider a circular resonator (often called a disk resonator) having one external port coupled through a capacitive gap, as shown in Fig. 2.14. The necessity of the coupling gap is first discussed.

When a planar circuit is used as a resonating element (for example, in a filter), it is usually desirable to keep the Q factor high. However, it is not easy when the external circuit is connected directly. For example, in the resonator described at the end of Sect. 2.5, $Q_0 = 910$ and $G = 0.27$ mS; hence the paral-

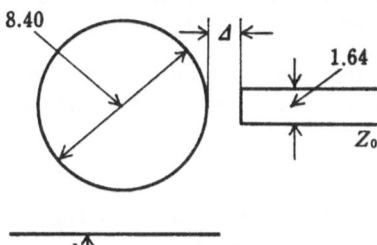

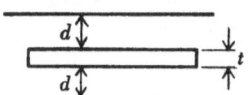

Fig. 2.14. An one-port resonator having a capacitive coupling. The dimensions of the model used in the experiment are also shown. All units are mm. Other parameters are $d = 1.45$ mm, $t = 7.2$ μm, $\varepsilon_s = 2.62$, $\tan \delta = 0.0014$ (at 10 GHz)

lel resistance $R = G^{-1} = 3.6\,k\Omega$. If we connect an external stripline having characteristic impedance $Z_0 = 50\,\Omega$, the loaded Q which is given as

$$Q_L = (Q_0^{-1} + Q_{ext}^{-1})^{-1}, \quad \text{where} \tag{2.77a}$$

$$Q_{ext}^{-1} = Q_0(Z_0/R), \tag{2.77b}$$

will drastically be lowered to about 13. Such a low Q value is useless in most cases.

An ordinary solution to this problem is to use a capacitive gap. When the reactance of the gap $(\omega C_c)^{-1}$ is much greater than Z_0, the external circuit including the gap seen from the resonator is approximately given as a parallel circuit of a capacitance C_c and a resistance $1/Z_0(\omega C_c)^2$. This resistance value can arbitrarily be raised by decreasing C_c, and an arbitrarily high Q_L value can thus be attained. Incidentally, in planar resonators for semiconductor microwave oscillators, the capacitive gap is sometimes useful also as a dc-current stopper.

The equivalent circuit of the capacitive gap was shown in the literature for various stripline-stripline couplings [2.10]. However, for planar-circuit-stripline couplings no detailed data are available; the electrostatic field analysis would become three dimensional and tedious. Therefore, only some typical data [Ref. 2.1, pp. 35 – 37] giving relations between the gap width Δ

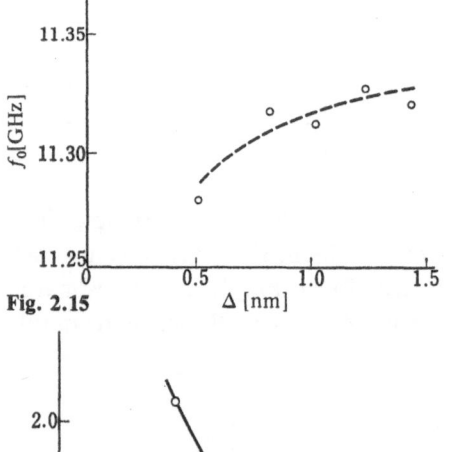

Fig. 2.15

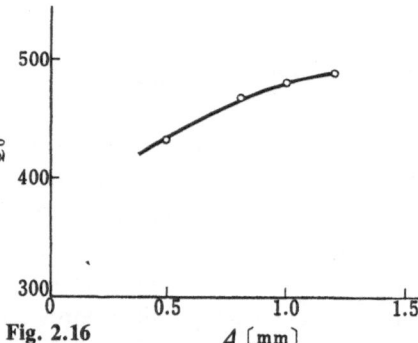

Fig. 2.16

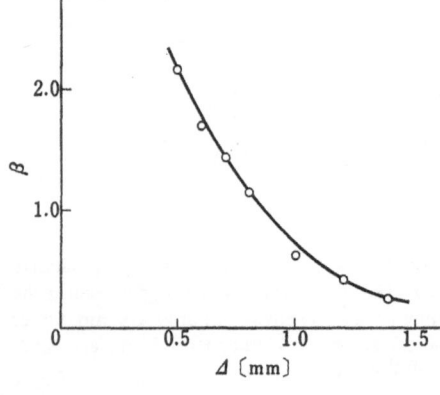

Fig. 2.15. Relation between the coupling-gap width and the resonant frequency [2.1]

Fig. 2.16. Relation between the coupling-gap width and the unloaded-Q factor. Q_0 (computed) = 547 [2.1]

Fig. 2.17. Relation between the coupling-gap width and the coupling coefficient β ($\beta = Q_0/Q_{ext}$) [2.1]

(Fig. 2.14), f_0 (resonant frequency), Q_0 (unloaded Q), and the coupling coefficient $\beta = Q_0/Q_{ext}$, where Q_{ext} is the external Q factor defined as $Q_{ext} = \omega C/$(external conductance), will be described in the following for the fundamental dipole mode ($m = n = 1$).

The parameters of the one-port circular resonator used in the experiment are shown in Fig. 2.14 and its caption. Figures 2.15 – 17 show f_0, Q_0, and β for the fundamental dipole mode, all as functions of the gap width Δ [Ref. 2.1, pp. 35 – 37]. The measurement has been performed by using a microwave network analyzer. The unloaded Q should be constant regardless of Δ; however, actually some variation is observed. [The formula $Q_0 = (Q_c^{-1} + Q_d^{-1})^{-1}$ gives $Q_0 = 547$.] Figures 2.15 – 17 tell us that Q_{ext} increases rapidly as Δ increases, whereas f_0 increases rather moderately.

The validity of (2.75) giving the correction for the fringe field has been investigated by comparing the theoretical and experimental resonant frequencies for the fundamental mode in two resonators having radii of 4.20 mm and 9.80 mm [Ref. 2.1, pp. 33 – 34]. The comparison of the experiment and the "corrected theory" indicates that in these particular cases, accuracy of 1% for the resonant frequency is assured by the correction of the fringe effect.

2.8.2 Coupled-Mode Filter Using a Single Circular Resonator

Generally, a planar circuit offers a more versatile performance than a stripline circuit does. A coupled-mode filter discussed in this section is an example [2.11, 12].

a) Theory

We consider two-port circular resonators as shown in Figs. 2.18a, b. Only the fundamental dipole mode ($m = n = 1$) is assumed to be excited.

The Green's function analysis can be applied to these circuits. For simplicity we assume that the widths of the ports are much less than the wavelength, and hence the factor F shown in Table 2.1 can be approximated as unity. Then

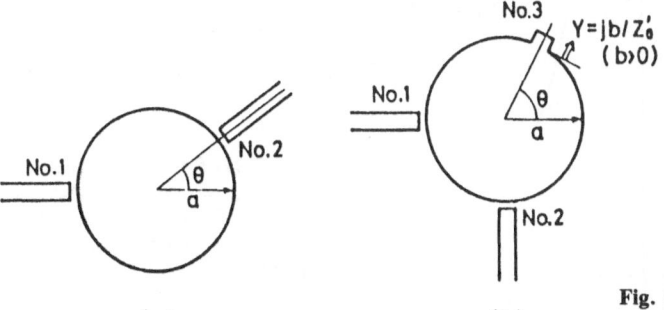

(a)   (b)

Fig. 2.18. Coupled-mode circular resonators [2.11]

the impedance matrix of the two-port resonator as shown in Fig. 2.18a is given, from (2.31, 44), as

$$\begin{pmatrix} Z_{11} & Z_{12} \\ Z_{21} & Z_{22} \end{pmatrix} = \begin{pmatrix} 1/Y_{11} & \cos\theta/Y_{11} \\ \cos\theta/Y_{11} & 1/Y_{11} \end{pmatrix},$$ (2.78)

where $Y_{11} (= G_{11} + jB_{11})$ denotes the input admittance of a one-port resonator of the same size and with the same gap.

For simplicity, we assume that both the source impedance and the load impedance are equal to a pure resistance Z_0', which is much higher than the characteristic impedance Z_0 of the line (see the third paragraph of Sect. 2.8.1). We further write

$$Y_{11}Z_0' = y_{11} = g_{11} + jb_{11}, \qquad Y_{in}Z_0' = y_{in}.$$ (2.79)

Then we obtain

$$y_{in} = y_{11} + \frac{\cos^2\theta}{1 + (1 - \cos^2\theta)/y_{11}}.$$ (2.80)

Therefore, for example, when $\theta = \pi$, $y_{in} = y_{11} + 1$ holds. When $\theta = \pi/2$, $y_{in} = y_{11}$ holds.

The power transmission can be calculated, by using the relation $S_{12} = 2Z_0 Z_{12}/[(Z_{11} + Z_0')(Z_{22} + Z_0') - Z_{12}^2]$ (S_{12}: element of the scattering matrix), as

$$|S_{12}|^2 = 4\cos^2\theta \frac{g_{11}^2 + b_{11}^2}{[(1 + g_{11})^2 - b_{11}^2 - \cos^2\theta]^2 + 4b_{11}^2(1 + g_{11})^2}.$$ (2.81)

This is illustrated in Fig. 2.19 as a function of the frequency. Not that unless $\theta = \pi$ or $\theta = \pi/2$, double-tuned characteristics are always found.

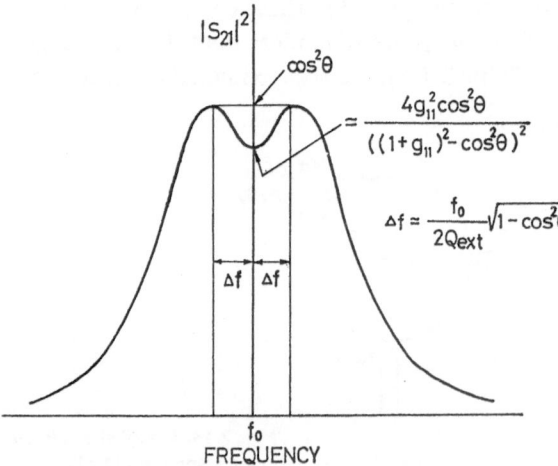

Fig. 2.19. Power transmission of a coupled-mode circular resonator. The parameter Q_{ext} is ordinarily defined: $Q_{ext} = Z_0\omega_0 C$ [2.12]

In this type of filter, however, the coupling is dependent on θ and is difficult to readjust after the circuit is completed. This difficulty is overcome by employing the structure of Fig. 2.18b. In this filter, Ports 1 (input) and 2 (output) are coupled to the horizontal and vertical dipole modes, respectively, which are originally decoupled. To give a necessary coupling between these two orthogonal modes, a coupling perturber having a small capacitive susceptance jb/Z_0' $(b > 0)$ is attached to Port 3.

In this case, if we assume for simplicity that the resonator itself is lossless $(y_{11} = jb_{11})$, the power transmission is given as

$$|S_{12}|^2 = \frac{4b^2\sin^2\theta\cos^2\theta}{[1 - b_{11}(b_{11} + b)]^2 + (2b_{11} + b)^2}. \tag{2.82}$$

This is shown for cases $b = 1.0$, 2.0, and 3.0 in Fig. 2.20. Equation (2.82) shows that the critical coupling is obtained for $b = 2.0$. The regions $b < 2.0$ and $b > 2.0$ correspond to the undercoupled and overcoupled conditions, respectively. Eq. (2.82) also shows that the maximum power transmission is obtained when a coupling perturber is attached to the position $\theta = \pi/4$.

b) Experiment

The experiment of the coupled-mode filter having a perturber has been performed in 11-GHz band [2.12]. Figure 2.21 shows the dimensions of the center conductor. This conductor is sandwiched by two ground conductors separated by Rexolite spacers (1.45 mm thick). The characteristic impedance of both the input and output striplines is 50 Ω.

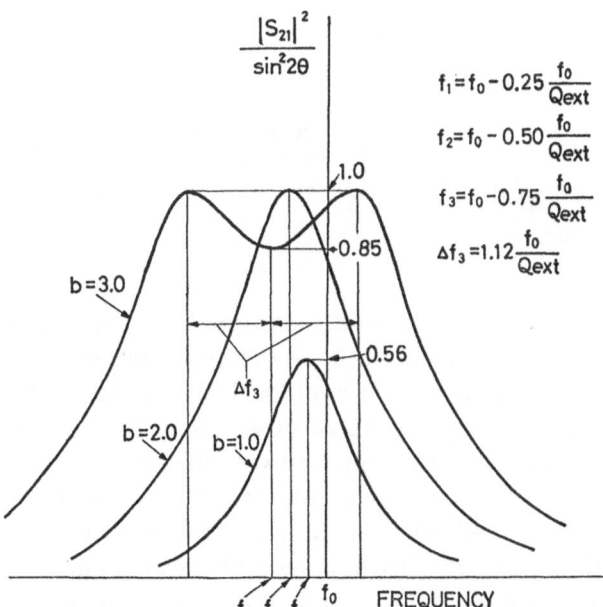

$$f_1 = f_0 - 0.25\frac{f_0}{Q_{ext}}$$

$$f_2 = f_0 - 0.50\frac{f_0}{Q_{ext}}$$

$$f_3 = f_0 - 0.75\frac{f_0}{Q_{ext}}$$

$$\Delta f_3 = 1.12\frac{f_0}{Q_{ext}}$$

Fig. 2.20. Power transmission of a coupled-mode circular resonator with a perturber for cases $b = 1.0$, 2.0, and 3.0. The parameter Q_{ext} is ordinarily defined: $Q_{ext} = Z_0\omega_0 C$ [2.12]

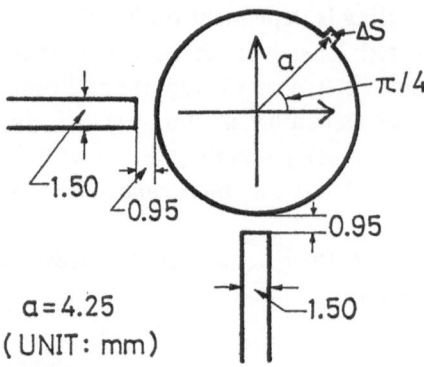

Fig. 2.21. Dimensions of the center conductor of an experimental coupled-mode filter [2.12]

a = 4.25
(UNIT: mm)

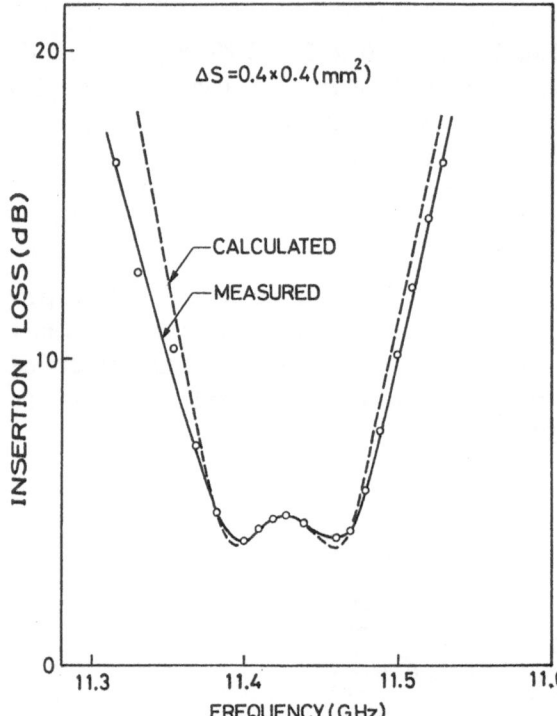

Fig. 2.22. Measured and calculated insertion loss of an experimental coupled-mode filter [2.12]

Figure 2.22 shows the measured and calculated insertion loss of the filter for the case when the size of the coupling perturber is 0.4×0.4 mm^2; this size gives a slight overcoupling. In this figure, a constant loss [dB] is added to the calculated loss so that the calculated and measured losses agree at the center frequency. A reasonably good agreement is found between the two curves.

Figure 2.23 shows the coupling coefficient of this filter, defined as $K = M/L$, where M and L are the equivalent circuit parameters shown in the same figure. The calculated coupling coefficient is shown by a solid line as a function of the area of the coupling perturber. The measured coupling coeffi-

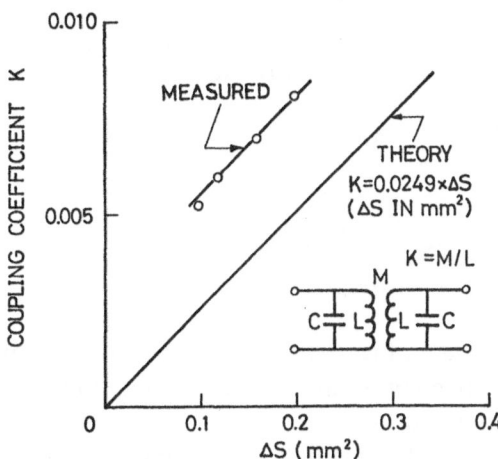

Fig. 2.23. Coupling coefficient versus the area of the coupling perturber [2.12]

cients are found along a somewhat separated line. The difference between the theory and experiment suggests that some initial coupling is present even in the absence of the coupling perturber.

2.8.3 Planar 3-dB Hybrid Using a Circular Resonator

The 3-dB hybrid plays an important role in microwave integrated circuits. The principal 3-dB hybrids have been the branch-line and ratrace circuits shown in Fig. 2.24. However, as circuit integration advances into the millimeter-wave region, the hybrid circuits based on the transmission-line concept encounter manufacturing difficulties. In view of this future trend, the possibility of realizing a 3-dB hybrid with simpler structure has been investigated [2.13, 14]. The outcome was a planar hybrid having four ports at angular spacings of 90°, 45°, 90°, and 135° around the periphery of a circular resonator.

a) Principle of the Planar 3-dB Hybrid

We consider a circular resonator having four ports (Fig. 2.25a). Under excitation at Port 1, the voltage at position (r, θ) is given as

$$V_{mn}(r, \theta) = V_0 \cos m\theta \, J_m(k'_{mn} r) . \tag{2.83}$$

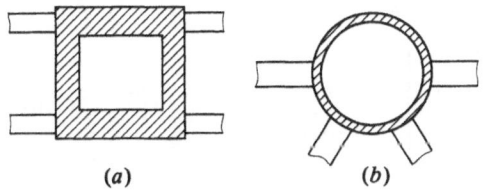

Fig. 2.24a, b. Distributed-constant 3-dB hybrids: (a) branch-line circuit, (b) ratrace circuit

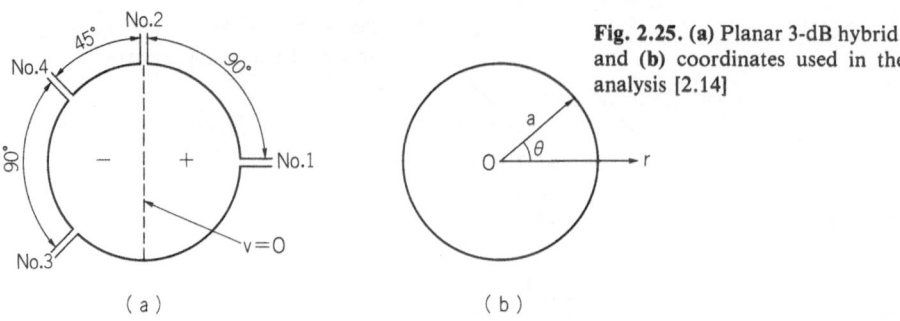

Fig. 2.25. (a) Planar 3-dB hybrid, and (b) coordinates used in the analysis [2.14]

(a)   (b)

Here the coordinate (r, θ) is defined as in Fig. 2.25b; k'_{mn} is the nth root of $J'_m(ka) = 0$, where a prime denotes differentiation; V_0 is a constant; and J_m is the mth−order Bessel function of the first kind.

The fundamental mode ($m = 1$, $n = 1$) is a dipole mode characterized by the plus and minus signs and the dashed line denoting the position corresponding to $V = 0$ in Fig. 2.25a. Therefore, no voltage is induced at Port 2 and voltages opposite in phase to that at Port 1 are induced at Ports 3 and 4. That is, couplings of

Port 1 → Port 3 (opposite phase) and Port 4 (opposite phase)

are obtained. When Ports 2, 3, and 4 are excited, similarly, couplings of

Port 2 → Port 3 (opposite phase) and 4 (same phase),
Port 3 → Port 1 (opposite phase) and 2 (opposite phase),
Port 4 → Port 1 (opposite phase) and 2 (same phase)

will take place, respectively. Hence the circuit in Fig. 2.25a is expected to be a hybrid.

To investigate the usefulness of this circuit as a hybrid, its frequency characteristics must be calculated, rigorously taking the effect of the higher-order modes into account. Such calculation is shown in the next section.

b) Impedance and Scattering Matrices

From (2.31, 44), the impedance matrix of the circuit shown in Fig. 2.25a is given as

$$Z_{ij} = \sum_{m=0}\sum_{n=1} \frac{j\omega\mu d \cos m(\theta_i - \theta_j)}{\varepsilon_m \pi (a^2 k_{mn}^2 - m^2)\left(1 - \dfrac{k^2}{k_{mn}^2}\right)}$$

$$\times \left[\frac{\sin\left(\dfrac{mW_i}{2a}\right)}{\dfrac{mW_i}{2a}}\frac{\sin\left(\dfrac{mW_j}{2a}\right)}{\dfrac{mW_j}{2a}}\right] + \frac{1}{j\omega C_0}. \tag{2.84}$$

The term in [ ] represents the effect of port width and can be approximated as 1 when $mW_i/2a \ll 1$ and $mW_j/2a \ll 1$ hold. Symbol C_0 denotes the electrostatic capacitance ($= 2\varepsilon\pi a^2/d$) of the circuit plate.

The scattering matrix can be determined from the above impedance matrix as

$$[S] = [\sqrt{Y_0}]([Z] - [Z_0])([Z] + [Z_0])^{-1}[\sqrt{Z_0}] , \qquad (2.85)$$

where

$$[Z_0] = \begin{bmatrix} Z_{01} & & & 0 \\ & Z_{02} & & \\ & & \ddots & \\ 0 & & & Z_{0N} \end{bmatrix} , \quad [\sqrt{Z_0}] = \begin{bmatrix} \sqrt{Z_{01}} & & & 0 \\ & \sqrt{Z_{02}} & & \\ & & \ddots & \\ 0 & & & \sqrt{Z_{0N}} \end{bmatrix} , \qquad (2.86a, b)$$

$$[\sqrt{Y_0}] = \begin{bmatrix} 1/\sqrt{Z_{01}} & & & \\ & 1/\sqrt{Z_{02}} & & \\ & & \ddots & \\ & & & 1/\sqrt{Z_{0N}} \end{bmatrix} , \qquad (2.86c)$$

and $Z_{01}, Z_{02}, \ldots, Z_{0N}$ denote the characteristic impedances of transmission lines connected to circuit ports. Since S_{ij} is more useful than $[Z_{ij}]$ in the discussion of microwave circuits, hereafter $[S_{ij}]$ will be considered.

c) Experiment

In [2.13, 14], experiments of five prototype circuits using Rexolite 1422 as the spacer were described. Here two examples of them will be introduced.

In one circuit (called here A), a (radius) = 9.46 mm, a' (radius corrected for fringe effect) = 10.13 mm, f_0 (dipole mode) = 5.46 GHz, and four 50 Ω striplines are connected with angular spacings of 90°, 45°, 90°, and 135°. In the other circuit (called here B), resonator dimensions are the same as A, but the coupled striplines are narrowed, as shown in Fig. 2.26, so as to reduce the possible adverse effect of higher-order modes.

The measured scattering parameters are shown by small crosses and dots in Fig. 2.27, whereas the dashed curves show theoretical ones. The good agreement obtained between theory and experiment seems to substantiate the validity of the principle of the planar hybrid and the above analysis. However, comparison with conventional hybrid circuits (ratrace type, for example) suggests that the bandwidth of the planar hybrid with respect to S parameters (except S_{12}) is relatively narrow [2.13, 14]. The only distinguished feature of the planar hybrid is its simple structure.

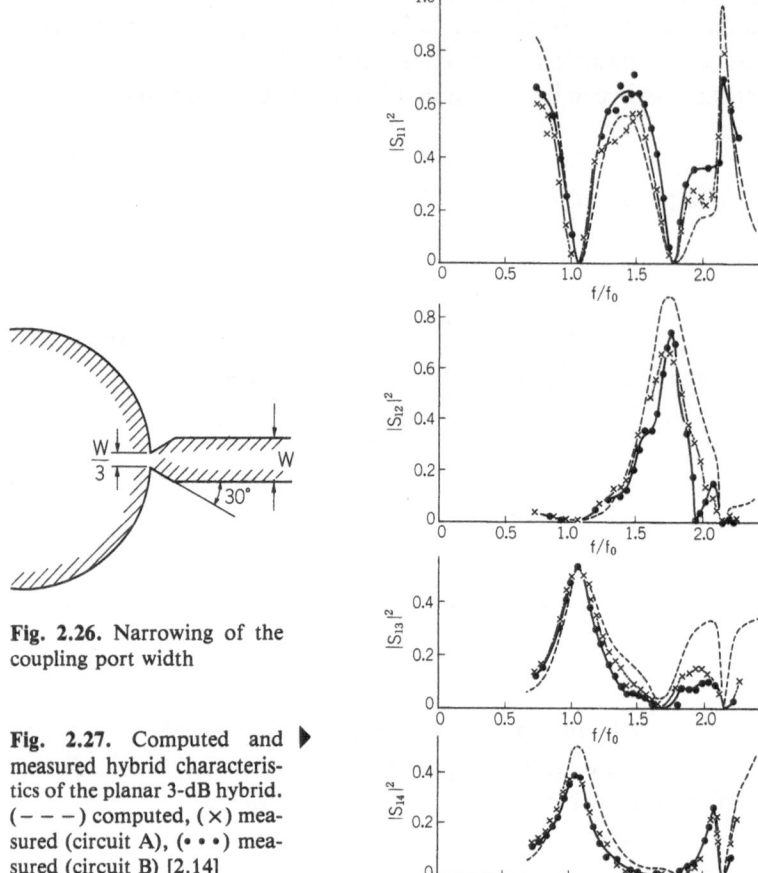

Fig. 2.26. Narrowing of the coupling port width

Fig. 2.27. Computed and ▶ measured hybrid characteristics of the planar 3-dB hybrid. (– – –) computed, (×) measured (circuit A), (• • •) measured (circuit B) [2.14]

2.9 Summary

The analysis of triplate-type planar circuits having simple shapes, based upon the Green's function of the wave equation, has been described. The analysis has been applied to three practical circuits to be used in MICs. The analysis of open-boundary planar circuits having arbitrary shapes is presented in the following chapter.

The literature cited so far in this chapter does not cover all the published papers dealing with planar circuits having simple shapes. Some papers from the earliest stage of the research have been mentioned in Chap. 1. Several other papers have also dealt with the problem after 1978 [2.15 – 17].

3. Analysis of Planar Circuits Having Arbitrary Shapes

In contrast to the preceding chapter, we here discuss the analysis of triplate-type planar circuits having arbitrary shapes. All the methods described in the present chapter are naturally computer-oriented.

In the major part of this chapter, the contour-integral method (also called the boundary-integral method), developed originally for planar circuit analysis, is presented. In this method, the wave equation is first converted to an integral equation along the circuit periphery. The equivalent circuit parameters are then derived from this contour-integral equation. Thus the required computer time is reduced appreciably as compared with other methods in which the field must be solved over the entire area of the circuit. A few examples of actual analyses using the contour-integral method are presented; more examples of its application will be vound in the following chapters.

In Sect. 3.7, analysis methods based upon the eigenfunction expansion are described. In these methods, conventional mathematical techniques such as the Rayleigh-Ritz method or finite-element method are used to obtain the eigenvalues and associated eigenfunctions, from which the Green's function is derived. Equivalent circuit parameters are then given in terms of the Green's function in the same manner as in the case of circuits having simple shapes.

3.1 Background

The numerical analyses presented in this chapter aim at the determination of the circuit parameters of triplate-type multiport planar circuits having arbitrary shapes in a wide frequency range.

In the preceding chapter, it has been shown that for planar circuits having simple shapes, the Green's functions can be obtained analytically, and circuit parameters are readily derived from these Green's functions. It has also been shown that when the circuit characteristics only in a narrow frequency range around a resonance point are to be obtained, an approximate, simple analysis based upon an energy consideration has been possible.

However, as emphasized in Chap. 1, one of the principal features of the planar circuit is that we can analyze and design an arbitrarily shaped circuit within a reasonable computer time. Practically, therefore, the analysis of an arbitrarily shaped circuit is technically much more significant. Such analysis will inevitably be computer-oriented.

Even for the arbitrarily shaped circuits, the computation of the eigenvalues and eigenfunctions, and derivation of the Green's function from these, are still possible if not easy. Conventional mathematical techniques such as the Rayleigh-Ritz method or finite-element method can be used to solve the eigenvalue problem. However, in these methods the eigenfunctions over the entire area of the circuit pattern must be computed; such computation is tedious and requires long computer time.

What we are primarily concerned with are the voltage and current distributions along the periphery. Therefore, a more direct analysis technique called the contour-integral method[1] has been developed specifically for planar circuit analysis [3.1]. In this method, the wave equation is first converted into an integral equation along the circuit periphery, and the circuit parameters are derived from this contour-integral equation. Thus the required computer time can be reduced appreciably. Besides, this method is equally applicable to an arbitrary circuit pattern; i.e., the computer time does not increase due to its complexity.

In the major part of this chapter, the description will be concentrated upon the contour-integral method. The analysis methods based upon eigenfunction expansion are described relatively briefly in Sect. 3.7. Throughout this chapter the circuit loss is apparently not considered. However, it can be taken into account, if necessary, simply by making the wavenumber k a complex quantity, as we did in Sect. 2.2.1.

3.2 Basic Formulation of the Contour-Integral Method

We consider an arbitrarily shaped, triplate planar circuit having several coupling ports (Fig. 3.1). Solving the wave equation (2.4a) over the entire area D inside the circuit periphery C will require a long computer time. However, when we are concerned only with the rf voltage along the periphery, such a computation is not necessary. The wave equation can be converted into a contour-integral form, which relates the voltage and current along the circuit periphery [3.1].

Using Weber's solution for cylindrical waves [3.2], the potential at a point upon the periphery is found to satisfy the following integral equation (see Appendix A 3.1 for the details of the derivation):

$$V(s) = \frac{1}{2j} \oint_C [k \cos \theta H_1^{(2)}(kr) V(s_0) - j \omega \mu d H_0^{(2)}(kr) i_n(s_0)] ds_0. \qquad (3.1)$$

---

[1] This method is also called the "boundary-integral method."

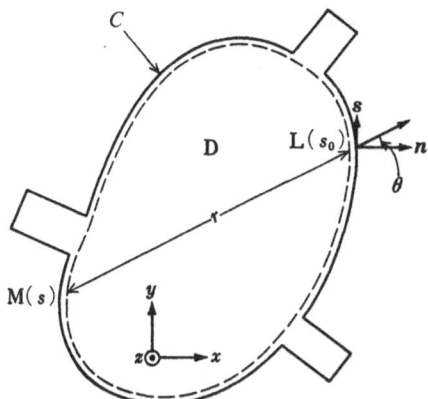

Fig. 3.1. Symbols used in the integral-equation representation of the wave equation

Here $H_0^{(2)}$ and $H_1^{(2)}$ are the zero-order and first-order Hankel functions of the second kind, respectively; i_n denotes the current density flowing outwards along the periphery; and s and s_0 denote the distance along Contour C. The variable r denotes distance between Points M and L represented by s and s_0, respectively, and θ denotes the angle made by the straight line from Point M to Point L and the normal at Point L, as shown in Fig. 3.1. Note that the function $H_0^{(2)}(kr)\exp(j\omega t)$ denotes a cylindrical wave traveling "outward" in a two-dimensional free space [3.3].

Equation (3.1) gives the relation between the rf-voltage and rf-current distributions along the periphery. When $i_n(s_0)$ is given, this equation is a second-kind Fredholm integral equation in terms of the rf voltage. However, in the present problem, i_n is given in turn in terms of $V(s_0)$, as seen in the following.

3.3 Circuit Parameters of an Equivalent N Port

To solve (3.1) numerically, we divide here the circuit periphery into N incremental sections numbered as 1, 2, ..., N, having widths $W_1, W_2, \ldots, W_N$, respectively, as illustrated in Fig. 3.2. Coupling ports are assumed to occupy each one of those sections. [If necessary, we may assume that a port (or ports) occupy two or more sections. However, for simplicity it is not the case in Fig. 3.2 and in following discussions]. Further, we set N sampling points at the center of each section.

When we assume that the magnetic and electric field intensities are constant over each width of these sections, the above integral equation results in a system of matrix equations:

$$\sum_{j=1}^{N} u_{ij} V_j = \sum_{j=1}^{N} h_{ij} I_j \quad (i = 1, 2, \ldots, N),$$ (3.2)

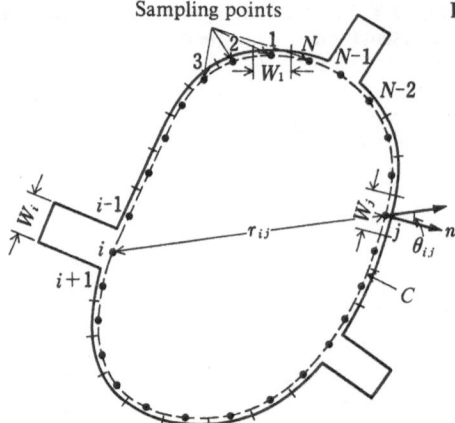

Sampling points   **Fig. 3.2.** Symbols used in the numerical analysis

where

$$u_{ij} = \delta_{ij} - \frac{k}{2j} \int_{W_j} \cos\theta \, H_1^{(2)}(kr) \, ds \,, \tag{3.3}$$

$$h_{ij} = \begin{cases} \dfrac{\omega\mu d}{4} \dfrac{1}{W_j} \int_{W_j} H_0^{(2)}(kr) \, ds & (i \neq j) \\[3mm] \dfrac{\omega\mu d}{4} \left[1 - \dfrac{2j}{\pi} \left(\ln \dfrac{kW_i}{4} - 1 + \gamma \right) \right] & (i = j) \,, \end{cases} \tag{3.4}$$

$\gamma = 0.5772 \ldots$ is Euler's constant, and $I_j = -2i_n W_j$ represents the total current flowing into the jth port on both the upper and lower surfaces of the circuit plate. The second formula in (3.4) has been derived by integrating the asymptotic expression of $H_0^{(2)}(kr)$ for $kr \ll 1$, assuming that the ith section is straight. The other formulas may be obtained in a straightforward manner. Equation (3.3) gives $u_{ii} = 1$ when W_i is small.

Solving (3.2), we can obtain the rf voltage on each sampling point as

$$V = U^{-1} HI \,. \tag{3.5}$$

Here V and I denote column vectors consisting of V_i and I_i, and U and H are $N \times N$ matrices consisting of u_{ij} and h_{ij}, respectively. The matrices U and H are given in terms of the circuit pattern and the manner of sectioning its periphery. The matrix U^{-1} is the inverse matrix to U.

We temporarily consider here that all the N sections upon the periphery are coupling ports and that the planar circuit is represented by an N-port equivalent circuit. Then, from the above relations, the impedance matrix of the equivalent N-port circuit is obtained as

$$Z = U^{-1} H \,. \tag{3.6a}$$

In actual numerical calculation, an matrix element Z_{ij} can be is given as

$$Z_{ij} = \frac{1}{\det U} \begin{vmatrix} u_{11} & \cdots & \overset{\displaystyle\downarrow i}{h_{1j}} & \cdots & u_{1N} \\ \vdots & & \vdots & & \vdots \\ u_{N1} & \cdots & h_{Nj} & \cdots & u_{NN} \end{vmatrix}, \tag{3.6b}$$

where $\downarrow i$ denotes substitution into the ith column.

In practice, however, most of the N ports described above are open-circuited. When external admittances are connected to several of them and the rest of the ports are left open-circuited, the reduced impedance matrix can be derived without difficulty, by constructing the impedance matrix by choosing only "necessary" elements given by (3.6b).

When all the N ports are open-circuited, i.e. when all I_j's are zero, (3.2) tells us that column vector $(V_1, V_2, \ldots, V_N)^t$ has a nontrivial (nonzero) solution only when

$$\det U = 0 \tag{3.7}$$

holds. The above equation gives the resonant frequency of such a completely open-circuited (i.e., isolated) planar circuit.

3.4 Transfer Parameters of a Two-Port Circuit

In the case of a two-port circuit (Fig. 3.3), the transfer parameters A, B, C, and D of the equivalent two-port circuit can be given more simply as follows. This method is essentially equivalent to the more general approach described above, but has a practical advantage in that the computation of an $N \times N$ inverse matrix required in (3.6a) is not necessary.

The circuit periphery is gain divided into N sections. Suppose P and Q denote the driving port and load port, respectively, as shown in Fig. 3.3. Admittances Y_p and Y_q are connected to those terminals, so that

$$\begin{aligned} Y_p &= 2i_n(p)\, W_p/V_p, \\ Y_q &= 2i_n(q)\, W_q/V_q. \end{aligned} \tag{3.8}$$

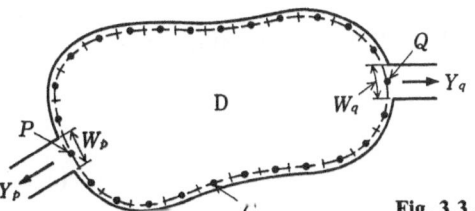

Fig. 3.3. An arbitrarily shaped two-port planar circuit

Here the voltages and current densities are assumed to be uniform over each section. Because an rf source is connected to the port P, Y_p has negative conductance component.

Equation (3.2) can be applied to all the N sampling points. Thus the rf voltage at each point can be given, from (3.8), by the following matrix equation:

$$(U + Y_p H_p + Y_q H_q) \begin{bmatrix} V_1 \\ V_2 \\ \vdots \\ V_N \end{bmatrix} = 0, \tag{3.9}$$

where H_p and H_q are again matrices determined by the shape of the circuit, defined in terms of matrix elements h_{ij} (3.4) as

$$H_p = \begin{bmatrix} & \overset{p}{h_{1p}} & \\ 0 & h_{2p} & 0 \\ & \vdots & \\ & h_{Np} & \end{bmatrix}, \tag{3.10}$$

$$H_q = \begin{bmatrix} & \overset{q}{h_{1q}} & \\ 0 & h_{2q} & 0 \\ & \vdots & \\ & h_{Nq} & \end{bmatrix}. \tag{3.11}$$

In order that a steady rf field exists in the circuit, from the nontrivial condition of $(V_1, V_2, \ldots, V_N)^t$ in (3.9),

$$\det [U + Y_p H_p + Y_q H_q] = 0 \tag{3.12}$$

must hold. This equation directly gives a bilinear relation between $-Y_p$, the driving point admittance, and Y_q, the load admittance, as

$$-Y_p = \frac{C' + D' Y_q}{A' + B' Y_q}, \tag{3.13}$$

where A', B', C', and D' are given as the following determinants:

$$A' = \det \begin{bmatrix} u_{11} & \overset{p}{h_{1p}} & u_{1N} \\ \vdots & \vdots & \vdots \\ u_{N1} & h_{np} & u_{NN} \end{bmatrix}, \tag{3.14}$$

$$B' = \det \begin{bmatrix} u_{11} & \overset{p}{h_{1p}} & \overset{q}{h_{1q}} & u_{1N} \\ \vdots & \vdots & \vdots & \vdots \\ u_{N1} & h_{Np} & h_{Nq} & u_{NN} \end{bmatrix}, \tag{3.15}$$

$$C' = \det [u_{ij}], \tag{3.16}$$

$$D' = \det \begin{bmatrix} u_{11} & \overset{q}{h_{1q}} & u_{1N} \\ \vdots & \vdots & \vdots \\ u_{N1} & h_{Nq} & u_{NN} \end{bmatrix}. \tag{3.17}$$

Equation (3.13) shows that A', B', C', and D' are quantities proportional to the so-called transfer parameters A, B, C, and D of the equivalent two-port circuit. In order that the reciprocity condition $(AD - BC = 1)$ holds, we should divide A', B', C', and D' by $(A'D' - B'C')$ to get A, B, C, and D, respectively, as

$$\begin{bmatrix} A & B \\ C & D \end{bmatrix} = \frac{1}{\sqrt{A'D' - B'C'}} \begin{bmatrix} A' & B' \\ C' & D' \end{bmatrix}. \tag{3.18}$$

When the circuit is a one-port circuit $(Y_q = 0)$, the input admittance is given simply as

$$-Y_p = C'/A'. \tag{3.19}$$

This equation also tells us that when the circuit has no coupling port $(Y_p = 0$, in addition to $Y_q = 0)$ and no circuit loss, $C' = 0$ gives the proper frequency, that is, the resonant frequency of the circuit. This has already been shown in the general formulation (3.7).

3.5 Numerical Computation Procedure

We next investigate the numerical computation procedure considering a two-port circuit as an example. In addition to the impedance matrix or transfer parameters calculated in the preceding section, other circuit characteristics such as the input admittance, power transmission, and rf-voltage distribution along the circuit periphery are also computed.

3.5.1 Description of Circuit Pattern

We first divide the periphery of the circuit pattern into N sections, and approximate these sections by straight-line elements (Fig. 3.4). These dividing

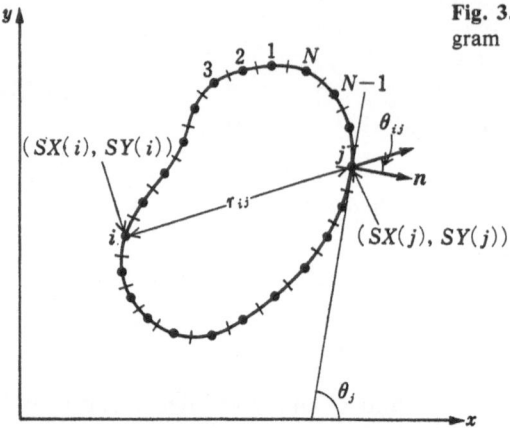

Fig. 3.4. Parameters used in the computer program

points are numbered $i = 1, 2, 3, \ldots$ in counterclockwise direction, and their coordinates are denoted as $(X(i), Y(i))$. The sections between the ith and $(i+1)$th dividing points are called ith sections, and sampling points are set at the center of each section, so that its coordinates $(SX(i), SY(i))$ are given as

$$SX(i) = \frac{X(i) + X(i+1)}{2}, \quad SY(i) = \frac{Y(i) + Y(i+1)}{2} \quad (i = 1, 2, \ldots, N).$$

$$(3.20)$$

The width of the ith section is given as

$$W_i = \sqrt{[X(i+1) - X(i)]^2 + [Y(i+1) - Y(i)]^2} .$$

$$(3.21)$$

When the circuit pattern has a hole or holes, the numbering of the sections along its periphery must be made in a opposite direction, i.e., it must be clockwise when the direction of the numbering along the outer periphery is counterclockwise. Thus, the computation is performed correctly regardless of the presence of the hole(s) because the direction outward normal to the periphery is always seen in the right-hand side along both the outer and inner peripheries of the circuit. This problem is discussed further in Appendix A4.1.

The widths of the sections need not be equal to each other. Usually it is desirable to use narrower sections at positions where the curvature is large, because the circuit periphery is approximated by a piecewise linear pattern.

3.5.2 Computation of r_{ij} and θ_{ij}

As shown in Figs. 3.2, 4, the distance between the ith and jth sampling points, and the angle made by the line connecting these points and the normal at the jth sampling points, are denoted as r_{ij} and θ_{ij}, respectively. The distance r_{ij} is given as

$$r_{ij} = \sqrt{[SX(i) - SX(j)]^2 + [SY(i) - SY(j)]^2} \,. \tag{3.22}$$

On the other hand, if we denote by θ_i the angle between the x axis and the line having a direction from point $(X(j), Y(j))$ to point $(X(j+1), Y(j+1))$, we may write

$$\cos \theta_{ij} = \{[SX(j) - SX(i)] \sin \theta_j - [SY(j) - SY(i)] \cos \theta_j\}/r_{ij} \,. \tag{3.23}$$

By using the relation

$$\cos \theta_j = [X(j+1) - X(j)]/W_j \,,$$
$$\sin \theta_j = [Y(j+1) - Y(j)]/W_j \,, \tag{3.24}$$

Eq. (3.23) can be rewritten as

$$\cos \theta_{ij} = \begin{cases} \{[SX(j) - SX(i)][Y(j+1) - Y(j)] - [SY(j) \\ \quad - SY(i)][X(j+1) - X(j)]\}/r_{ij} W_j \\ 0 \quad (i = j) \,. \end{cases} \tag{3.25}$$

Thus θ_{ij} can be computed from coordinates of sampling and dividing points.

3.5.3 Computation of Matrix Elements u_{ij} and h_{ij}

The matrix elements u_{ij} and h_{ij} defined in (3.3, 4) may be obtained precisely by computing the integrals using Simpson's method. However, in many cases these can be computed approximately, without computation of integrals, as

$$u_{ij} = \delta_{ij} - \frac{k}{2j} \cos \theta_{ij} H_1^{(2)}(kr_{ij}) W_j \,, \quad h_{ij} = \frac{\omega \mu d}{4} H_0^{(2)}(kr_{ij}) \,. \tag{3.26}$$

The choice between the above two alternative approaches should be made by considering the accuracy and computer time.

In actual circuit analyses, a major part of the computer time is used in the computation of Bessel functions. When the number of division is N, the total time required in the Bessel-function computation is proportional to N^2.

3.5.4 Computation of Transfer Parameters

The transfer parameters are computed as determinants [see (3.14 – 17)]. To compute large-scale determinants, the so-called sweep-out method is usually used. It is easily shown that the time required to compute an $N \times N$ determinant is proportional to $(N^2 - N)$, which can be approximated as N^2 when N is large. Thus, as a rule of thumb, the overall computation time (including the Bessel-function computation described above) is proportional to N^2. This fact limits the number of division N.

When N is not large enough, the computation error increases. The presence of the computation error is detected most simply by noting that A and D must be real, where as B and C must be imaginary, when the circuit is lossless. Moreover, when the circuit pattern is symmetrical and the input and output ports are provided at symmetrical positions,

$$A = D \tag{3.27}$$

must hold. This second fact is also often used to check the computation accuracy.

3.5.5 Computation of Input Admittance and S_{12}

All the circuit characteristics can be computed from the transfer parameters. Some basic relations follow.

The input admittance $-Y_p$ can be given, from (3.13, 18), as

$$-Y_p = (C + DY_q)/(A + BY_q) . \tag{3.28}$$

We consider first the matching conditions. We note here that A and D are real, whereas B and C are imaginary. Hence, in order to obtain a real Y_p for a real Y_q (load admittance), from (3.28),

$$Y_q^2 = AC/BD \tag{3.29}$$

must hold. In other words, the real Y_p for a real Y_q is obtained at those frequencies where (3.29) holds. Naturally, at such frequencies, substituting (3.29) into (3.28), we obtain

$$-Y_p = \frac{D}{A} Y_q . \tag{3.30}$$

When both the circuit shape and coupling-port positions are symmetrical, from (3.27), we have

$$-Y_p = Y_q . \tag{3.31}$$

This means, when both the input and output lines have a characteristic admittance Y_0 and the circuit is symmetrical, the impedance match is realized at frequencies where

$$Y_0 = C/B . \tag{3.32}$$

For a more general case when the input and output line impedances are Z_G and Z_L, respectively, a component S_{21} of the scattering matrix S is given as

$$S_{21} = \frac{2Z_L}{AZ_L + B + CZ_G Z_L + DZ_G} = T_r + jT_i. \tag{3.33}$$

Therefore, the insertion loss T and phase delay φ are

$$T = -10\log_{10}(|S_{21}|^2 Z_G / Z_L) \quad \text{[dB]}, \tag{3.34}$$

$$\varphi = -\tan^{-1}(T_i / T_r). \tag{3.35}$$

3.5.6 RF Voltage Along the Circuit Periphery

Once the input admittance $-Y_p$ is determined by the preceding process, the rf voltage distribution along the circuit periphery can readily be obtained by solving (3.9) as a simultaneous equation for V_i's. In the actual computation, we divide V_i's into real and imaginary components, and assume that at an appropriate sampling point $V_i = 1 + j0$. Thus, (3.9) becomes a set of simultaneous equations for $2(N-1)$ variables.

The entire circuit characteristics of a planar circuit having an arbitrary shape can be obtained by the process described in the preceding Sects. 3.5.1 – 5. As described in Sect. 3.7, there are several other methods for analyzing the arbitrarily-shaped planar circuit characteristics. However, the contour-integral approach is usually the most timesaving and versatile.

3.6 Examples of Computer Analysis by the Contour-Integral Method

3.6.1 One-Port Circular Circuit

As an example of the computer analysis, the input admittance of a one-port circular circuit with $\varepsilon_s = 2.62$, $a = 1.841$ m, $d = 0.628$ m is described first. These values are not realistic ones; $a = 1.841$ m is chosen so that the fundamental resonant wavenumber is given by $k = 1$ m^{-1}. Note that 1.841 is the first root of $J_1'(x) = 0$. The circuit is assumed to be lossless. Figure 3.5 shows the circuit pattern and symbols used in the analysis.

The result for the case $W = 0$ (Fig. 3.5) is shown in Fig. 3.6 [3.1]. This figure shows the variation of the input admittance, given by (C'/A'), around the fundamental resonant frequency. The parameter N denotes the number of the sampling points along the periphery. As N increases, the real frequency locus approaches the values obtained by the simple theory as described in the preceding chapter, shown by the small crosses along the ordinate in Fig. 3.6. Note that the abscissa is expanded by a factor of ten to exaggerate the computation error.

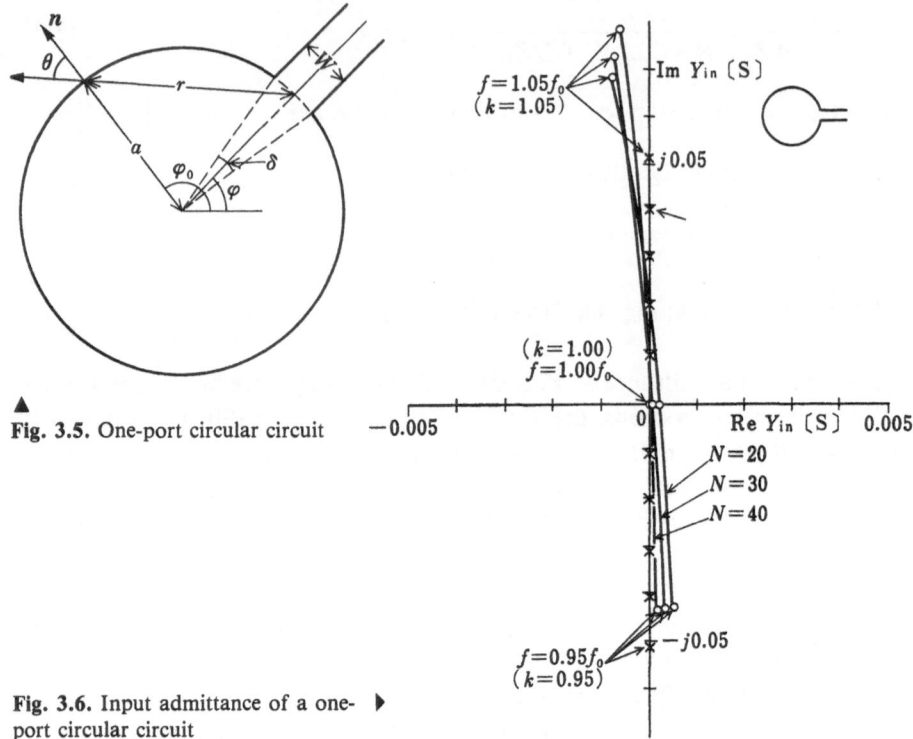

Fig. 3.5. One-port circular circuit

Fig. 3.6. Input admittance of a one- ▶ port circular circuit

The values of k giving $C' = \det U = 0$ correspond to the resonant frequencies of the circuit (3.7). From the simple analysis described in Sect. 2.4.2, they should satisfy $J'_m(ka) = 0$. For $a = 1.841$ m, k should then be 1.000 m^{-1}, 1.659 m^{-1}, and so forth. This fact provides a simple method for checking the computation accuracy.

Since C' is complex due to the computation error and $C' = 0$ is never realized for real k, we define the eigenvalue as k which gives the minimum of $|C'|$. The variation of $|C'|$ is shown as a function of k in Fig. 3.7, which shows the first ($k = 1.00$) and the second ($k = 1.66$) minima. The former corresponds to the fundamental dipole mode [the first root of $J'_1(ka) = 0$] and the latter to the quadrupole mode [the first root of $J'_2(ka) = 0$]. Table 3.1 shows

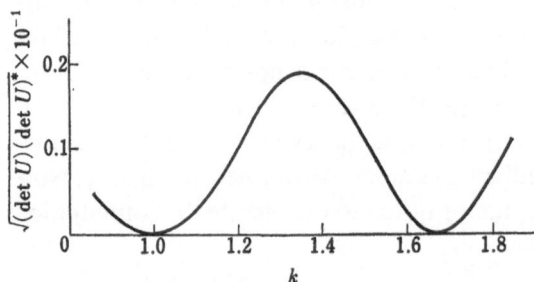

Fig. 3.7. The frequency dependence of $|\det U|$ for $N = 40$

Table 3.1. The first eigenvalue of the circular circuit described in the text

N (number of divisions)	k (computed eigenvalue)
20	1.00013
30	1.00008
40	1.00007

the first k obtained for various N. It is found that k tends toward unity as N increases.

3.6.2 Two-Port Circular Circuit

Next the transfer parameters A, B, C, and D of a circular circuit having two ports on its opposite sides computed by using (3.14 – 18) are shown in Fig. 3.8 [3.1]. In this figure the abscissa gives the real part and the ordinate the imaginary part of the transfer parameters obtained in the case $N = 40$. Parameters A and D are equal to each other as the circuit is symmetrical, and they are -1.0 at the resonant frequency.

By using the obtained transfer parameters and the relation $Y_{in} = (C + D/R_L)/(A + B/R_L)$, the input admittance of the circular circuit loaded by a pure resistance R_L is computed, as shown in Fig. 3.9. The curves show the computer calculations for load resistances of 50, 500, and 5000 Ω. The loci in this figure cover a frequency range from the dipole mode of resonance to the quadrupole mode of resonance. Between these two low-admittance, parallel resonant points we find the frequency where the input admittance is very high, that is, a series resonant point, on the right-hand side of each locus. Note that such a frequency can never be found except by computer analysis.

Figure 3.10 shows the rf-voltage distribution along the periphery for $R_L = 500$ Ω and $N = 40$ at various frequencies. In this figure, both ends and the center of the abscissa correspond to the load port and the driving port, respectively. The solid and broken curves show the magnitude (arbitrary scale) and phase of the rf voltage along the periphery, respectively. It is found that as the frequency increases, the distribution of the rf voltage changes from a dipole mode to a quadrupole mode. At the frequency of $1.12 f_0$, the rf voltage at the input port is minimized; this corresponds to the series resonance of the circuit.

Figure 3.11 shows the power transmission calculated numerically by using the relation $S_{21} = 2Z_0/(A Z_0 + B + C Z_0^2 + D Z_0)$ for the case when the charac-

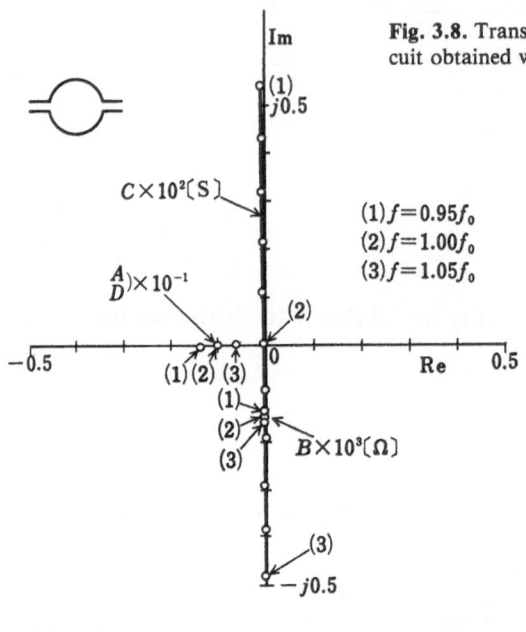

Fig. 3.8. Transfer parameters of a two-port circular circuit obtained with $N = 40$

$(1) f = 0.95 f_0$
$(2) f = 1.00 f_0$
$(3) f = 1.05 f_0$

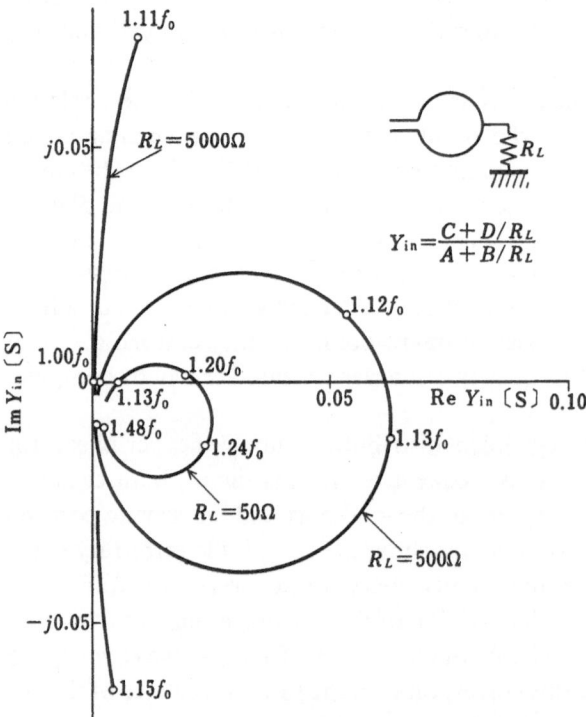

$$Y_{in} = \frac{C + D/R_L}{A + B/R_L}$$

Fig. 3.9. Input admittance of a two-port circular circuit obtained with $N = 40$

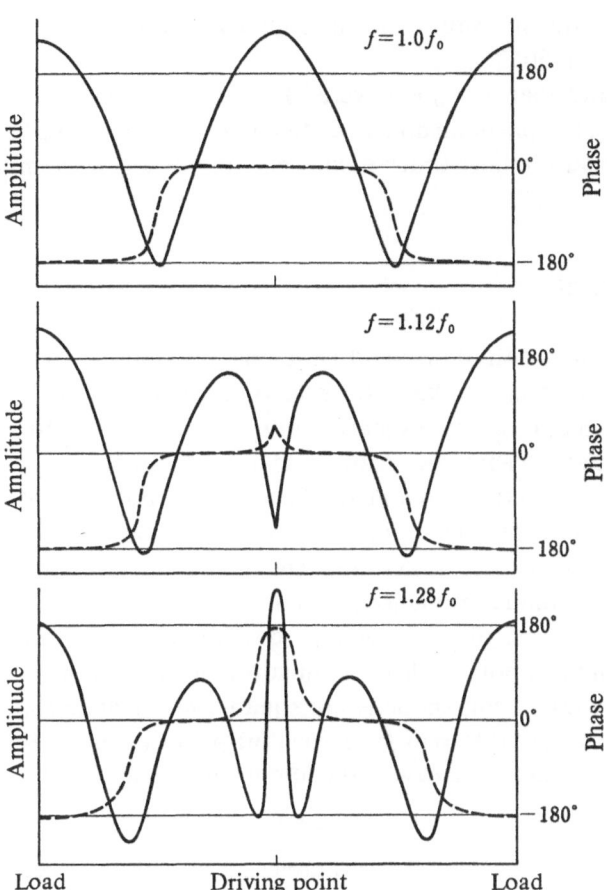

Fig. 3.10. The rf voltage distributions for various frequencies along the periphery of a two-port circular circuit. The solid and broken curves show the amplitude and phase, respectively. Both ends of the abscissa correspond to the position of the load terminal; $R_L = 500 \ \Omega$, $N = 40$

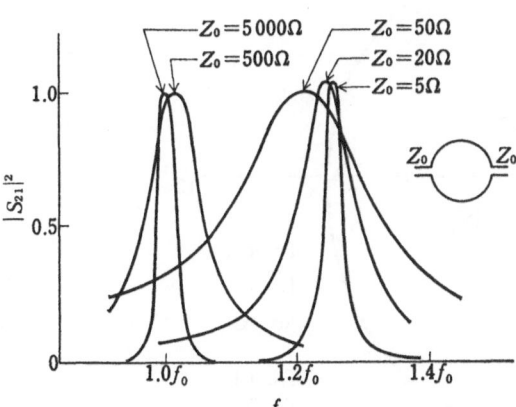

Fig. 3.11. Power transmission through a two-port circular circuit for various load impedances

teristic impedances Z_0 of the input and output lines are both pure resistances and equal to each other. It is found that both the frequency giving maximum power transmission and the transmission bandwidth increase as the line impedance Z_0 is lowered down to 50 Ω.

However, it is also found that as Z_0 is lowered further, the bandwidth becomes again narrowed. This phenomenon is difficult to explain using a simple physical picture, because higher-order modes play important roles in such cases.

3.6.3 Two-Port Square Circuit

Figures 3.12, 13 show the input admittance and rf-voltage distribution of a square circuit computed in a mannar similar to the preceding subsection [3.1]. The circuit pattern is assumed to be an $a \times a$ square with $\varepsilon_r = 2.62$, $a = 4.44$ m, and $d = 0.628$ m. These values are again not realistic as in Sect. 3.6.1.

Symbol f_0 in Fig. 3.12 denotes the resonant frequency of the quadrupole mode. When the frequency increases, the dipole-mode resonance occurs first at Point B; then the series resonance occurs at a frequency a little above $0.80f_0$ (Point C), and the quadrupole-mode resonance at Point A.

In Fig. 3.13, only the real part of the rf-voltage distribution is shown. However, the variation from the dipole mode to the quadrupole mode is clearly seen in this figure. When the circuit shape is not square but rectangular, having length and width somewhat different from each other, a similar analysis shows that, as the frequency increases, the dipole mode along the

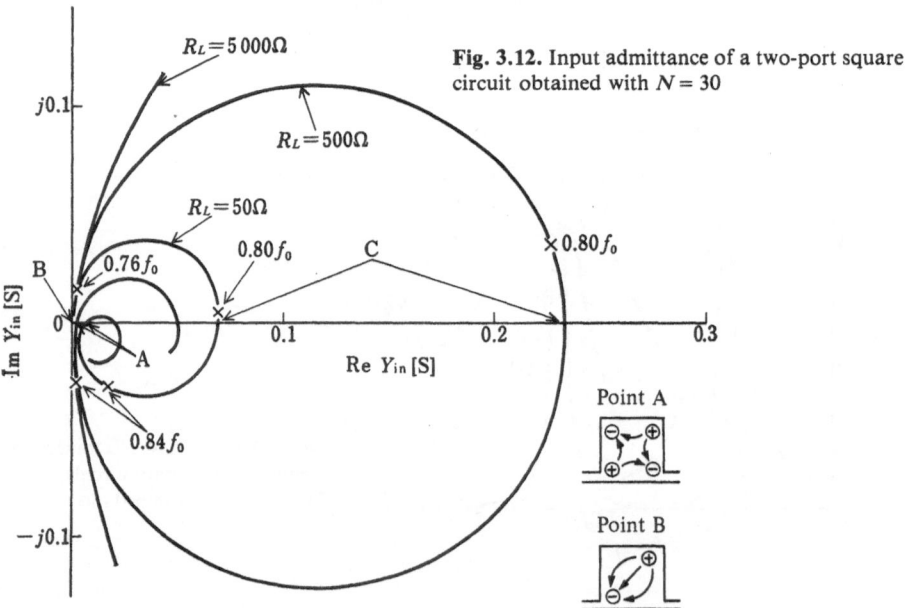

Fig. 3.12. Input admittance of a two-port square circuit obtained with $N = 30$

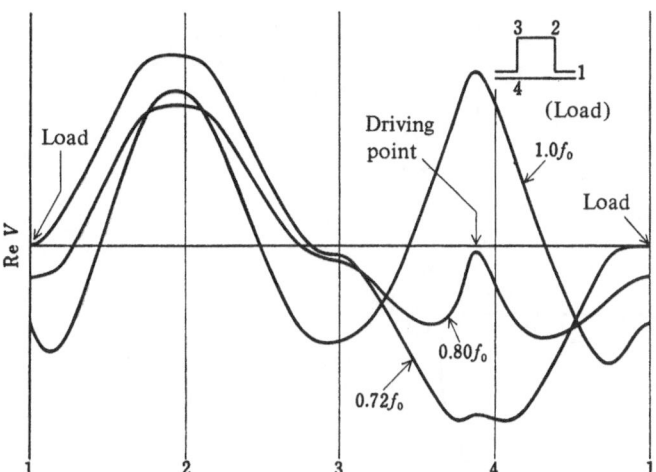

Fig. 3.13. The rf voltage distribution along the periphery of a two-port square circuit. Only the real part of the voltage phasor is shown; $R_L = 50\ \Omega$, $N = 32$

"length" occurs first, followed by another dipole mode along the "width." Finally, the quadrupole mode occurs. Of course, many higher modes follow further.

3.6.4 Irregularly Shaped Circuit

As an example of a more irregularly shaped circuit, the characteristics of a planar circuit as shown in the inset of Fig. 3.14, is given. Figure 3.14 illustrates the frequency loci of the input admittance for $R_L = 500\ \Omega$ computed with $N = 32$. In this figure, f_0 denotes the resonant frequency of the quadrupole mode of a regular square circuit having dimensions $2a \times 2a$. Parallel resonances are found at $0.54 f_0$ and $0.86 f_0$, and a series resonance at $0.64 f_0$.

3.6.5 Computer Time

In the above numerical computations, most of the computer time is consumed in the computations of the matrix elements and determinants. When the number of the divisions of the periphery is N, the computer time required is proportional to N^2 in the matrix-element computation, and to $N(N-1)$ in the determinant computation (Sects. 3.5.3, 4).

When a modern large-scale, high-speed computer is used, the determinant computation takes, for a 40×40 matrix, for example, typically 0.1 s, whereas the element computation usually takes a longer time. In the element computation, the computation of Bessel functions (more generally, Hankel functions) is predominant in time. Three methods can be used in the Hankel-function computation:

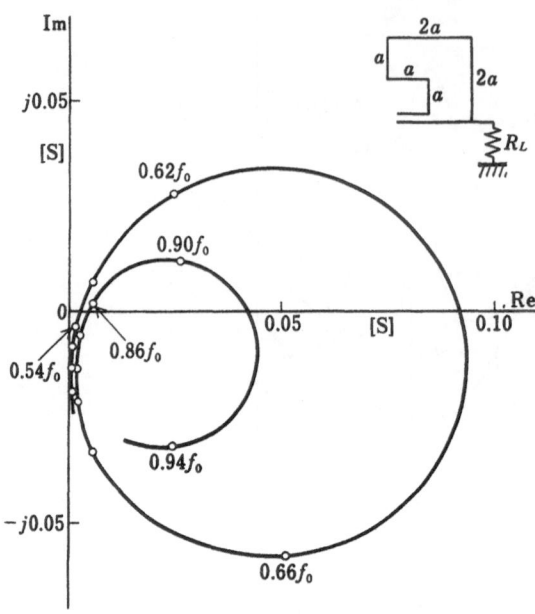

Fig. 3.14. Input admittance of an irregularly shaped circuit: $R_L = 500\ \Omega$, $N = 32$

1) the library program,
2) four-point interpolation (Lagrange's method) between Hankel-function values stored beforehand,
3) approximate polynomial expressions of Hankel functions.

All these methods have been tested; the required computer times were almost in the same order of magnitude when the same accuracy was to be assured [3.1].

3.7 Analyses Based on Eigenfunction Expansion

3.7.1 Advantages and Disadvantages of Eigenfunction-Expansion Approaches

There are several other methods for the analysis of an arbitrarily shaped triplate-type planar circuit, all of which are based upon eigenfunction expansion. In these methods, the analysis of the circuit characteristics follows the following three steps as in Chap. 2: (i) the computation of eigenfunctions and eigenvalues, (ii) the computation of the Green's function, and (iii) the computation of equivalent circuit parameters from the Green's function. When the circuit pattern is not simple, the above step (i) becomes usually very much complicated. However, once the eigenfunctions and eigenvalues are determined, steps (ii) and (iii) are essentially identical to the case of simply shaped circuits described in Sect. 2.3.

The eigenfunction-expansion methods are less versatile and in most cases more time-consuming than the contour-integral method because of the three steps required. Step (i) is particularly time-consuming. However, in some cases in which the circuit performance must be obtained for a large number of frequencies, it becomes advantageous (less time-consuming) because the frequency characteristics can easily be computed by changing the wavenumber k in (2.38). In contrast to this, with the contour-integral method, all the matrix elements must be newly computed when the frequency changes.

3.7.2 Methods for Solving Eigenvalue Problems

The problem to be solved here is the eigenvalue of a two-dimensional Helmholtz equation for an arbitrarily shaped open boundary. This is mathematically identical to the solution of TE-mode waves in a hollow metallic waveguide having an arbitrary cross section [3.4 – 10], and very similar to some aspects of the scattering problem of a metallic cylinder having an arbitrary cross section [3.11 – 13]. All to these methods described in [3.4 – 13] are more or less helpful in the present problem.

Principal techniques for solving the two-dimensional eigenvalue problems are classified as

1) the finite-difference method [3.5],
2) the point-matching method [3.8],
3) the Rayleigh-Ritz method [3.9, 10],
4) the finite-element method [3.6, 7].

In the finite-difference method, the field function is represented by values at latticelike sampling points, and the wave equation is solved as a simultaneous equation for these sampled field values. Usually, the two-dimensional relaxation method is used to solve the equation. However, this method is not easy to apply because the eigenfunction finally obtained is strongly influenced by the initial (starting) field pattern, and is scarcely used in the planar circuit analysis.

In the point-matching method, the field function is represented by a series consisting of terms of the form $A_n \cos(n\theta) J_n(k_n r)$ and/or $B_n \sin(n\theta)$ $\cdot J_n(k_n r)$, and parameters A_n, B_n, and k_n are determined from the boundary conditions. This method may be useful in the planar circuit analysis, but has not actually been used for this purpose.

The Rayleigh-Ritz method and the finite-element method are both based upon the variational principle. These have actually been applied to the analysis of the planar circuit, and will be described in some detail in Sects. 3.7.4 – 6.

3.7.3 Silvester's Theory

Before describing the two variational methods, *Silvester*'s theory [3.14], which was the earliest proposal of the eigenfunction-expansion analysis of planar circuits, is briefly introduced.

Silvester showed that the admittance matrix of a parallel-plate planar circuit as shown in Fig. 3.15 can be expressed as

$$Y = j \frac{\sqrt{\varepsilon/\mu}}{h} \sum_{k=0}^{\infty} \frac{\Omega}{\Omega_k^2 - \Omega^2} B^{(k)}, \quad \Omega = \omega\sqrt{\varepsilon\mu}, \qquad (3.36)$$

where h denotes the circuit height, and the (m, i) element of matrix $B^{(k)}$ is given as

$$B_{mi}^{(k)} = -\Omega_k^2 \frac{\int_R \psi_i \phi_k dR \int_R \psi_m \phi_k dR}{\int_R \phi_k^2 dR}, \quad B_{mi}^{(0)} = -\int_R \nabla\psi_i \cdot \nabla\psi_m dR. \quad (3.37)$$

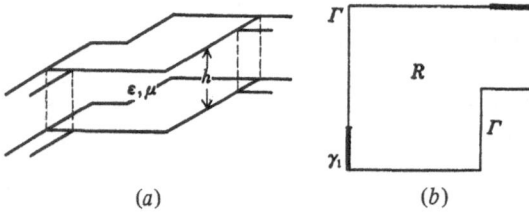

Fig. 3.15a, b. Circuit configuration and symbols used in Silvester's analysis: (a) Configuration of the planar circuit, (b) Symbols [3.14]

In the above equations, ψ_i denotes the solutions of

$$\nabla^2 u = 0 \quad (\text{in } R), \quad \partial u/\partial n = 0 \quad (\text{on } \Gamma),$$
$$u = 1 \quad (\text{on } \gamma_i), \quad u = 0 \quad (\text{on } \gamma_k, \ k \neq i), \qquad (3.38)$$

where γ_k denotes a part of the circuit periphery (Fig. 3.15b),

$$\nabla^2 = \partial^2/\partial x^2 + \partial^2/\partial y^2, \qquad (3.39)$$

and Ω_i and ϕ_i are the eigenvalues and eigenfunctions, respectively, of the following eigenvalue problem:

$$(\nabla^2 + \Omega^2) w = 0 \quad (\text{in } R), \quad \partial w/\partial n = 0 \quad (\text{on } \Gamma), \quad w = 0 \quad (\text{on } \gamma_k). \qquad (3.40)$$

The circuit characteristics can be obtained by solving (3.38, 40), computing (3.37), and substituting those results into (3.36).

However, in Silvester's method two eigenvalue problems (3.38, 40) must be solved, and hence it needs longer computer time than the method based upon (2.38). Nowadays Silvester's method is scarcely used, although it is important historically.

3.7.4 Solution of an Eigenvalue Problem by Variational Method

The Rayleigh-Ritz method and the finite-element method are both based upon the variational principle, which is described in the following rather briefly.
1) A quantity determined by the shape of a function $f(x)$ defined in a given range of the independent variable x is called a functional. In short, a functional is a function of functions.
2) A differential equation giving the condition to make a functional stationary (maximum or minimum) is called the Euler's equation for that functional.
3) When we solve a differential equation by using the variational method, we first find a functional whose Euler's equation is the given differential equation, and next we find the stationary condition of the functional. Thus the differential equation can be solved.

Here the problem to be solved is the Neumann (open-boundary) problem of a two-dimensional wave equation

$$(\nabla^2 + k^2)u = 0 . \tag{3.41}$$

In the following, it is shown that the above problem is equivalent to finding a stationary condition of a functional $I[u]$ defined, in Area D shown in Fig. 3.16, as

$$I[u] = \frac{J[u]}{K[u]} , \quad \text{where} \tag{3.42}$$

$$J[u] = \iint_D (u_x^2 + u_y^2)\, dx\, dy , \tag{3.43}$$

$$K[u] = \iint_D u^2\, dx\, dy , \tag{3.44}$$

and that the value of $I[u]$ thus determined is nothing but the eigenvalue k^2 to be obtained. In (3.43, 44), u_x and u_y denote $\partial u/\partial x$ and $\partial u/\partial y$, respectively.

The stationary condition means that the first-order variation of the functional δI is zero for a small change of the variable δu. Therefore, if at that point $I[u] = \lambda$ (a constant), then from (3.42) we may write

$$\delta J - \lambda \delta K = 0 , \quad \text{where} \tag{3.45}$$

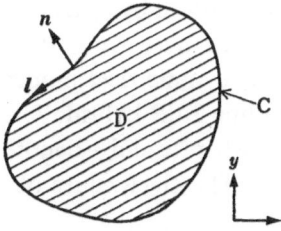

Fig. 3.16. Symbols used in the variational method analysis

$$\delta J = 2 \iint_D (u_x \delta u_x + u_y \delta u_y) \, dx \, dy \,, \tag{3.46}$$

$$\delta K = 2 \iint_D u \delta u \, dx \, dy \,. \tag{3.47}$$

Here we modify (3.46). By integration by parts, we have

$$\delta J = 2 \iint_D \left[\frac{\partial}{\partial x} (u_x \delta u) + \frac{\partial}{\partial y} (u_y \delta u) \right] dx \, dy - 2 \iint_D (u_{xx} + u_{yy}) \, \delta u \, dx \, dy$$

$$= 2 \oint_C [u_x \delta u \, dy + u_y \delta u \, dx] - 2 \iint_D (u_{xx} + u_{yy}) \, \delta u \, dx \, dy$$

$$= 2 \oint_C \frac{\partial u}{\partial n} \delta u \, ds - 2 \iint_D \nabla^2 u \delta u \, dx \, dy \,. \tag{3.48}$$

Substituting (3.47, 48) into (3.45), we obtain

$$0 = \oint_C \frac{\partial u}{\partial n} \delta u \, ds - \iint_D (\nabla^2 u + \lambda u) \, \delta u \, dx \, dy \,. \tag{3.49}$$

In (3.49), δu is an arbitrary function of position on C (the first term) and in D (the second term). Hence the two terms must both be zero. If the open-boundary condition is satisfied, the first term becomes zero, and vice versa. The second term has a form identical to the left-hand side of (3.41), provided that $\lambda = k^2$. Thus, the equivalence of the given problem with the variational problem, and that the stationary value of $I[u]$ gives k^2, have both been proved.

3.7.5 Rayleigh-Ritz Method

In the Rayleigh-Ritz method, an eigenfunction ϕ is expanded by a double series of base functions f_{mn} [such as $x^m y^n$, or $\cos(m\pi x) \cos(n\pi y)$, or $r^m \cos n\theta$] as

$$\phi = \sum_m \sum_n C_{mn} f_{mn} \,, \tag{3.50}$$

for which the coefficients C_{mn} are determined so as to make the functional I stationary.

Various problems relevant to planar circuits have been solved by *Hsu* et al. using this method [3.15, 16]. Only one example is cited here from their achievements. Figure 3.17 shows the lowest nine doubly symmetrical modes (eigenfunctions) in a planar structure used in a branch-line 3-dB hybrid. The cosine-type base functions [$f_{mn} = \cos(m\pi x) \cos(m\pi y)$, where x and y denote normalized coordinates] have been used in the analysis. The eigenvalues obtained are also shown in the figure.

$n=2$ $k=2.2076$ $n=3$ $k=2.5994$ $n=4$ $k=4.0784$

$n=5$ $k=5.4054$ $n=6$ $k=6.6404$ $n=7$ $k=6.8314$

$n=8$ $k=7.4117$ $n=9$ $k=7.8200$ $n=10$ $k=9.6970$

Fig. 3.17. The lowest nine doubly symmetrical eigenfunctions in a branch-line 3-dB hybrid [3.15]

3.7.6 Finite-Element Method

In the finite-element method, Area D inside the circuit pattern (Fig. 3.16) is first divided into a large number of triangular "elements" (typically $30-300$). The eigenfunction in each of such elements is then approximated, generally speaking, by a polynomial of the form $\phi = \sum k_{mn}x^m y^n$. However, in most applications of the finite-element method, only the lowest three terms are considered, that is, $\phi = ax + by + c$. This means that the eigenfunction is approximated by a linear function of (x, y) in one triangular element. Such an approximate function has a degree of freedom of 3 (a, b, and c), and hence can be determined by the ϕ values at three apexes of the element. When the entire Area D is covered by such elements, ϕ is approximated by a "piecewise-flat" function of (x, y) determined by the ϕ values at all the apexes in D. Finally, such ϕ values (ϕ_1, ϕ_2, ... ϕ_N, where N is the number of apexes) are determined so as to make the functional I stationary.

A number of papers describing the finite-element method analysis have appeared. Here only one example will be introduced from one of the earliest papers by *Miyoshi* and *Sakakibara* [3.17]. In this example, not only the eigenfunction expansion but also the circuit characteristics of a triplate-type T-branch circuit are derived.

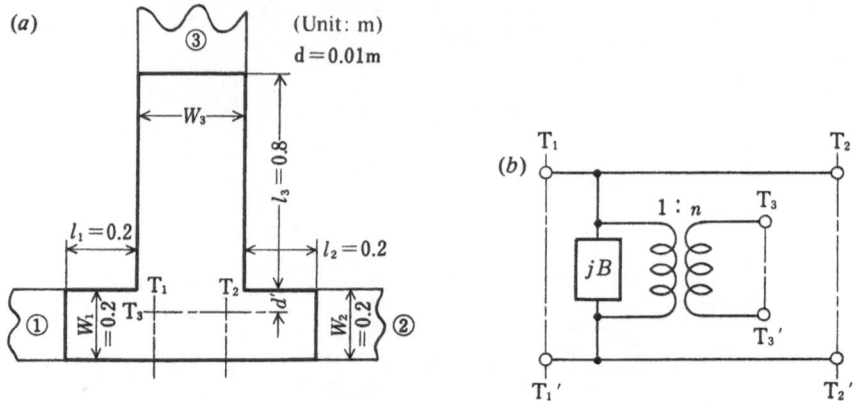

Fig. 3.18. (a) A T-branch circuit. (b) Equivalent circuit of the T-branch circuit shown in (a)

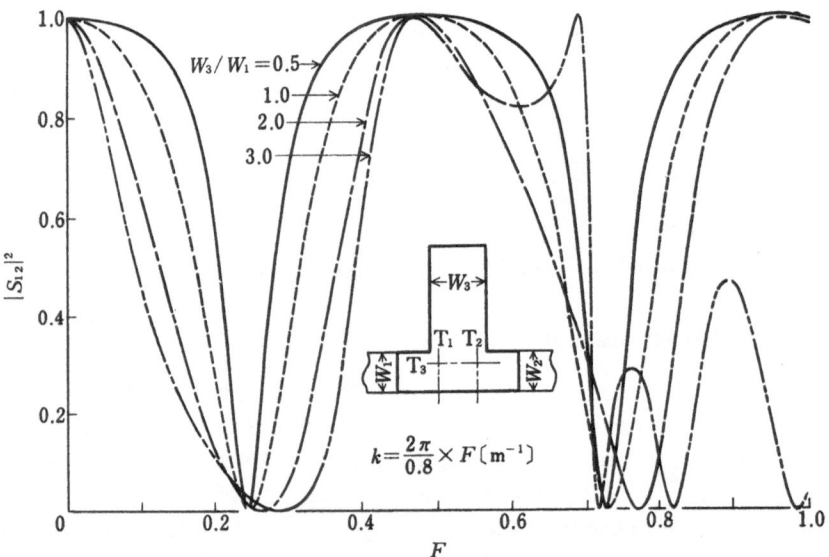

Fig. 3.19. Frequency characteristics of the power transmission ratio of the T-branch circuit with Port 3 open circuited, computed by the finite-element method [3.17]

The circuit to be considered is shown in Fig. 3.18a. Port 3 is assumed to be open circuited. After computing the eigenfunctions and associated eigenvalues of the T-shaped area, the power transmission ratio from Port 1 to Port 2, $|S_{12}|^2$ has been computed for various ratios (W_3/W_1) as functions of frequency. Figure 3.19 shows the result. It is seen that series resonances of the branch occur at frequencies where l_3 approaches $(2n+1)\lambda/4$ where λ denotes wavelength, making $|S_{12}|^2$ zero. It is also found that when (W_3/W_1) increases, (i) the cutoff bandwidth increases, and (ii) the stopping frequencies are raised a little.

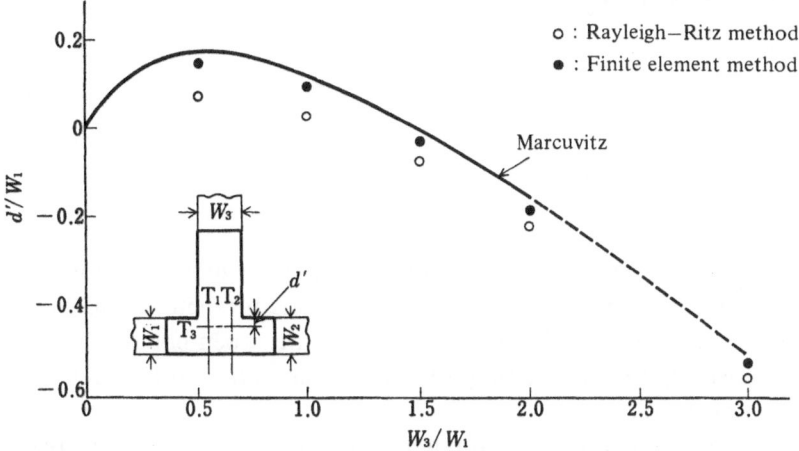

Fig. 3.20. Position of the reference plane T_3 as a function of width W_3 [3.17]

The bandwidth increase can be qualitatively understood from the fact that the branch impedance is lowered as W_3 increases. The second effect is more difficult to understand intuitively, but has been well known by circuit engineers. *Marcuvitz* reported an approximate analysis of this effect [3.18] and showed the shift of the reference plane T_3 (Fig. 3.19) which is defined so that length $(l + d')$ gives the quarter wavelength of the lowest center stopping frequency. In other words, the reference plane T_3 corresponds to the port $T_3 - T_3'$ of the lumped-constant equivalent circuit of the T junction shown in Fig. 3.18b.

Figure 3.20 shows the normalized position of the T_3 plane (d'/W_1) as functions of (W_3/W_1). We compare the result of Marcuvitz's approximate analysis, data from the finite-element method analysis (Fig. 3.19), and the result of a Rayleigh-Ritz method analysis performed for the comparison [3.17]. In the finite-element method analysis, the number of variables is 70, and 66 eigenfunctions are computed. In the Rayleigh-Ritz method analysis, the eigenfunctions are approximated by polynomials of maximum power 10, and the number of variables is 66 [3.17]. Good agreement is found among the three results, especially between Marcuvitz's analysis and the finite-element method.

3.8 Summary

Analyses of triplate-type planar circuits having arbitrary shapes have been described. The methods of analysis can be classified into two major categories: the contour-integral approach and those based on the eigenfunction expansion. In this chapter, in addition to the contour-integral method which has been most emphasized, two other approaches (Rayleigh-Ritz and finite-element methods) have been described together with some results of these analyses.

4. Short-Boundary Planar Circuits

A short-boundary planar circuit is essentially a thin waveguide circuit in which no transverse electric field exists. Circuits of this category require a mathematical formulation a little different from that of the open-boundary planar circuits because of the Dirichlet-type boundary condition and different input/output structures.

In this chapter a method of computer analysis of short-boundary planar circuits having arbitrary shapes is presented. The method is based upon the contour-integral representation of the two-dimensional wave equation. Results of the computer analyses for simple circuits are compared with analytical solutions to show the validity and accuracy of the proposed method of analysis. Some examples of analysis of practical circuits are also presented.

4.1 Background

As described in Sect. 1.2.1, the planar circuit can be classified into three types: the triplate type, open (or asymmetrical) type, and short-boundary type.

In the preceding two chapters, we have concentrated upon the analysis of the triplate-type planar circuits, which have open-circuit boundaries. Practically, the investigation of the triplate type is particularly important in connection with the microwave IC design.

In this chapter, the method for computer analysis of arbitrarily shaped, short-boundary planar circuits is described, which is technically significant in the design of a class of waveguide circuitry. Some examples of analyses of practical circuits are also presented.

The computer-analysis technique described in this chapter enables us to know the precise characteristics of circuits such as are shown in Fig. 4.1a – c. Moreover, in the case of Fig. 4.1b, the height of the waveguide need not be small compared to the wavelength as required by the definition of the planar circuit; the present analysis can also be applied to the ordinary TE_{10}-mode waveguide circuitry, provided that no transverse electric field is present. Therefore, the advantages of the computer-analysis technique of the short-boundary planar circuit extend to conventional waveguide technology.

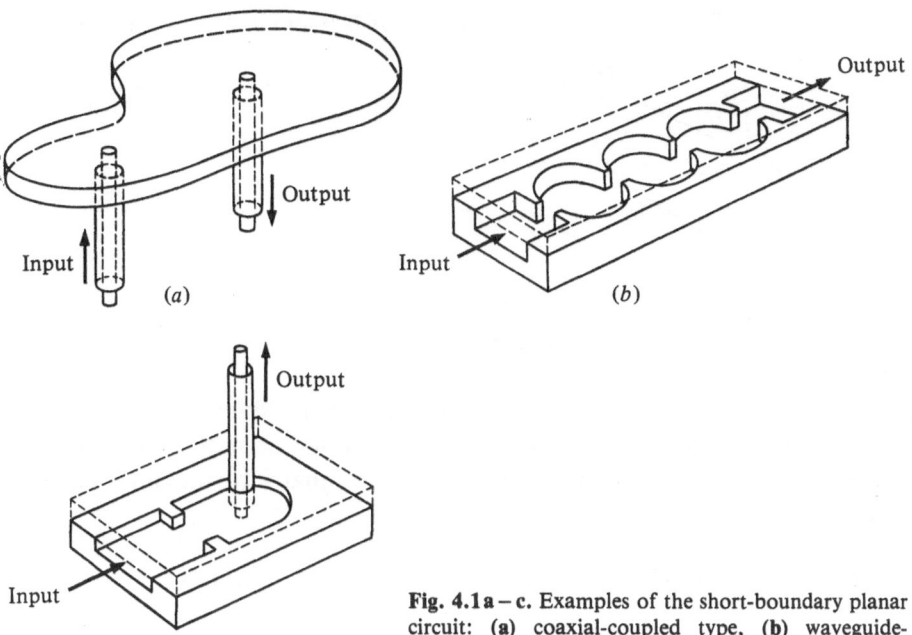

Fig. 4.1a – c. Examples of the short-boundary planar circuit: (**a**) coaxial-coupled type, (**b**) waveguide-coupled type, (**c**) mixed type

The analysis method described in this chapter is a contour-integral method [4.1] which is similar to that applied to open-boundary planar circuits in the preceding chapter [4.2]. The contour-integral method is not the only method applicable to the short-boundary planar circuits; other methods such as the eigenfunction-expansion method [4.3] and finite-element method [4.4, 5] are also applicable. However, these methods require relatively complicated formulation of the input/output conditions, and hence are omitted here because of space limitations.

4.2 Principle of Analysis

The computer-analysis technique developed for triplate-type planar circuits may be modified to its "dual" form for application to the short-boundary planar circuits. The most important modification stems from the fact that the coupling ports are of entirely different form. For example, when a planar circuit is coupled to the external circuits through waveguides (Fig. 4.1b), a computational process is required to provide the "match" between the electromagnetic field in the planar circuit and the proper fields in the waveguides at appropriately selected reference planes. The coaxial ports shown in Fig. 4.1a, c also require a similar computational process. In any case, however, the

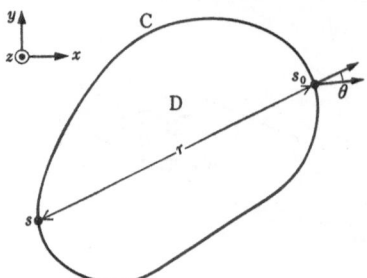

Fig. 4.2. Symbols used in the basic equation

basic equation for the open-boundary planar circuit can be utilized in the earlier stage of the analysis.

In Sect. 3.2, it was shown that by using Weber's solution of the two-dimensional wave equation, the rf voltage V at a point upon the periphery C of an arbitrarily shaped, homogeneous two-dimensional wave medium (Fig. 4.2) is given as

$$2j\,V(s) = \oint_C [k\cos\theta\,\mathrm{H}_1^{(2)}(kr)\,V(s_0) - j\,\omega\mu\,d\,\mathrm{H}_0^{(2)}(kr)\,i_n(s_0)]\,ds_0\,. \tag{4.1}$$

Here, $\mathrm{H}_0^{(2)}$ and $\mathrm{H}_1^{(2)}$ are the zero-order and first-order Hankel functions of the second kind, respectively, i_n denotes the current density flowing outwards along the periphery, s and s_0 denote the distance along the periphery C. The variable r denotes the distance between Points s and s_0, and θ denotes the angle made by the straight line from Point s to Point s_0, and the normal at Point s_0, as shown in Fig. 4.2. When i_n is given, (4.1) is a second-kind Fredholm integral equation with respect to the rf voltage V.

To make the description comprehensive, we restrict the following discussion to two cases. The first case is the short-boundary planar circuit having two coaxial coupling ports (Fig. 4.1a). The second case is that having two waveguide coupling ports (Fig. 4.1b). The description of the latter case will be emphasized for its practical importance. More complicated circuits such as that shown in Fig. 4.1c and those circuits having three or more ports can be dealt with by modifying or combining the analyses for the previous two cases.

4.3 Short-Boundary Planar Circuit Having Two Coaxial Coupling Ports

4.3.1 Basic Equation

For numerical calculation we divide the periphery of a circuit as shown in Fig. 4.1a into M incremental sections, and provide M sampling points at the center of each section (Fig. 4.3). We assume that current flows uniformly in each

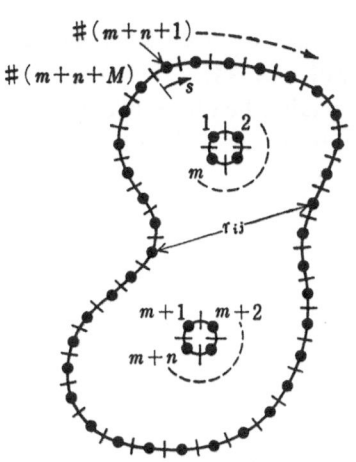

$\#(m+n+1)$

$\#(m+n+M)$

s

$1 \quad 2$

m

r_{ij}

$m+1 \quad m+2$

$m+n$

Fig. 4.3. Symbols used in the computer analysis – I
(case of the coaxial-coupled type)

section. The peripheries of the coupling conductors are also divided into m and n incremental sections, and sampling points are provided. These $(m+n+M)$ sampling points are numbered as follows:

1) Conductor 1: $i = 1 \sim m$;
2) Conductor 2: $i = (m+1) \sim (m+n)$;
3) Circuit periphery: $i = (m+n+1) \sim (m+n+M)$.

It was shown in Sect. 3.3 that if we rewrite (4.1) in an incremental form, we obtain a matrix equation

$$\sum_{j=1}^{N} u_{ij}V_j = \sum_{j=1}^{N} h_{ij}I_j \quad (i = 1, 2, \ldots, N), \quad \text{where} \tag{4.2}$$

$$u_{ij} = \delta_{ij} - \frac{k}{2j} \int_{W_j} \cos\theta \, H_1^{(2)}(kr) \, ds, \tag{4.3}$$

$$h_{ij} = \begin{cases} \dfrac{\omega\mu d}{2} \cdot \dfrac{1}{W_j} \int_{W_i} H_0^{(2)}(kr) \, ds & (i \neq j) \\[4mm] \dfrac{\omega\mu d}{2} \left[1 - \dfrac{2j}{\pi} \left(\ln\dfrac{kW_i}{4} - 1 + \gamma \right) \right] & (i = j). \end{cases} \tag{4.4}$$

In the above equations, γ denotes Euler's constant ($= 0.5772$); W_i, W_j, and I_j ($= -i_n W_j$) denote the widths of Port i and Port j, and the current flowing into Port j, respectively; and $N \triangleq m+n+M$. The difference by the factor of 2 between (3.4) and (4.4) stems from the fact that in the triplate case (3.4), the current is given as $I_j = -2i_n W_j$ because it flows on both the upper and lower surfaces of the circuit plate.

Equation (4.1) is derived in Sect. 3.3 for the case in which there is no "hole" in the circuit, in other words, in which the circuit pattern is singly connected. However, (4.1) is applicable also to a multiply connected pattern such as is shown in Fig. 4.3; the proof is given in appendix A4.1.

In actual computations, an approximate formula like (3.26) can often be used with enough accuracy. In the present case, however, a slightly modified form must be used for h_{ij}:

$$h_{ij} = \frac{\omega\mu d}{2} H_0^{(2)}(kr_{ij}) , \tag{4.5}$$

where the coefficient is doubled.

4.3.2 Simplification of the Basic Equation

For simplicity we assume the following conditions for the position and size of the coupling conductors:
1) The radius of the conductors R is much less than the wavelength ($R \ll \lambda$ or $kR \ll 1$).
2) If we denote the distance from the center of the conductor to the nearest spot upon the circuit periphery by r_{min}, $R \ll r_{min}$ holds.

Then we may assume that the voltage V and current density i_n are both uniform along the periphery of the conductor. Therefore, if the voltages and currents of two terminals p and q (Fig. 4.3) are denoted by V_p, V_q, I_p, and I_q, respectively, the voltages and currents in each section around the conductors are given as V_p, V_q, I_p/m, and I_q/n.

[We may remove the above assumption if we consider $(m-1)$ and $(n-1)$ higher-order modes in coaxial waveguides 1 and 2, respectively. Then we may take into account the (probably) reactive line impedances for those higher-order modes, and we no longer need the reduction of the constraints performed to obtain (4.7, 8). The formulation of such an analysis will be shown in Sect. 4.6 in connection with the input/output boundary conditions in a waveguide-type circuit.]

On the other hand, $V = 0$ holds along the periphery of the circuit. Therefore, we may write

$$[U] \begin{bmatrix} V_p \\ \vdots \\ V_p \\ \hline V_q \\ \vdots \\ V_q \\ \hline 0 \\ \vdots \\ 0 \end{bmatrix} \begin{matrix} \Big\} m \\ \\ \Big\} n \\ \\ \Big\} M \end{matrix} = [H] \begin{bmatrix} I_p/m \\ \vdots \\ I_p/m \\ \hline I_q/n \\ \vdots \\ I_q/n \\ \hline I_{m+n+1} \\ \vdots \\ I_{m+n+M} \end{bmatrix} \begin{matrix} \Big\} m \\ \\ \Big\} n \\ \\ \Big\} M \end{matrix} . \tag{4.6}$$

Equation (4.6) consists of $N(=m+n+M)$ scalar equations, whose number is greater than the number of variables $(M+2)$. To decrease the number of constraints, we define reduced matrices with $(M+2)\times(M+2)$ elements:

$$
[U'] =
\begin{bmatrix}
\displaystyle\sum_{i=1}^{m}\sum_{j=1}^{m} u_{ij} & \displaystyle\sum_{i=1}^{m}\sum_{j=m+1}^{m+n} u_{ij} & \vdots & \displaystyle\sum_{i=1}^{m} u_{i(m+n+1)}\cdots \\[2ex]
\displaystyle\sum_{i=m+1}^{m+n}\sum_{j=1}^{m} u_{ij} & \displaystyle\sum_{i=m+1}^{m+n}\sum_{j=m+1}^{m+n} u_{ij} & \vdots & \displaystyle\sum_{i=m+1}^{m+n} u_{i(m+n+1)}\cdots \\[2ex]
\hline
\displaystyle\sum_{j=1}^{m} u_{(m+n+1)j} & \displaystyle\sum_{j=m+1}^{m+n} u_{(m+n+1)j} & \vdots & u_{m+n+1 \atop m+n+1} \cdots \\[2ex]
\vdots & \vdots & & \cdots\; u_{m+n+M \atop m+n+M}
\end{bmatrix}
\tag{4.7}
$$

$$[H'] = \text{(similar to the above)} . \tag{4.8}$$

Then we may further rewrite (4.6) as

$$
[U']
\begin{bmatrix}
V_p \\ V_q \\ \hline 0 \\ \vdots \\ 0
\end{bmatrix}
\begin{matrix} \Big\}2 \\ \\ \Big\}M \end{matrix}
= [H']
\begin{bmatrix}
I_p/m \\ I_q/n \\ \hline I_{m+n+1} \\ \vdots \\ I_{m+n+M}
\end{bmatrix}
\begin{matrix} \Big\}2 \\ \\ \Big\}M \end{matrix}
\tag{4.9}
$$

The previous simplification implies that *each of the m (or n) sampling points is equally weighted.*

4.3.3 Derivation of Admittance Parameters

We may derive the admittance parameters Y_{pp}, Y_{pq}, Y_{qp}, and Y_{qq} directly from H' and U', as we did in Sect. 3.3 for the impedance parameters.

First, we temporarily consider that all the $(M+2)$ sampling points are coupling terminals and that the planar circuit is represented by an $(M+2)$-port equivalent circuit. The admittance matrix Y of such a circuit is given from (4.9) as

$$Y = H'^{-1}U' . \tag{4.10}$$

The wanted parameters Y_{pp}, Y_{pq}, Y_{qp}, and Y_{qq} will be found in the top left corner of the matrix Y. This method can readily be applied to cases in which the circuit has three or more ports.

Practically, however, the previous computation requires rather long computer time. When the circuit has only two ports, we have a simpler alternative, which is described in the following subsection.

4.3.4 Derivation of Transfer Parameters

We assume that the terminals p and q are driving and load terminals, respectively, and impedances Z_p and Z_q are connected to them. Then Z_p must have a negative real part, and must be equal to the driving point impedance multiplied by -1, provided that a stable rf field exists in the circuit. Since

$$Z_p = -V_p/I_p, \tag{4.11}$$

$$Z_q = -V_q/I_q \tag{4.12}$$

hold, (4.9) is rewritten as

$$[H' + mZ_pU_p + nZ_qU_q] \begin{bmatrix} I_p/m \\ I_q/n \\ I_{m+n+1} \\ \vdots \\ I_{m+n+M} \end{bmatrix} = 0, \tag{4.13}$$

where U_p and U_q are again matrices determined by the shape of the circuit

$$U_p = \begin{bmatrix} \overset{1}{u'_{11}} & 0 \cdots 0 \\ u'_{21} & \vdots \quad \vdots \\ \vdots & \vdots \quad \vdots \\ u'_{(M+2)1} & 0 \cdots 0 \end{bmatrix}, \tag{4.14a}$$

$$U_q = \begin{bmatrix} 0 & \overset{2}{u'_{12}} & 0 \cdots 0 \\ \vdots & u'_{22} & \vdots \quad \vdots \\ \vdots & \vdots & \vdots \quad \vdots \\ 0 & u'_{(M+2)2} & 0 \cdots 0 \end{bmatrix}. \tag{4.14b}$$

For (4.13) to have a nontrivial solution, i.e., a steady field in the circuit,

$$\det [H' + mZ_pU_p + nZ_qU_q] = 0 \tag{4.15}$$

must hold. This equation leads directly to a bilinear relation between $-Z_p$, the driving point impedance, and Z_q, the load impedance, as

$$-Z_p = \frac{A'Z_q + B'}{C'Z_q + D'},$$ (4.16)

where A', B', C', and D' are given as the following determinants:

$$A' = n \det \begin{bmatrix} h'_{11} & u'_{12} & h'_{13} & \cdots & h'_{1N'} \\ h'_{21} & u'_{22} & & & \vdots \\ \vdots & \vdots & \vdots & \vdots & \vdots \\ h'_{N'1} & u'_{N'2} & h'_{N'3} & \cdots & h'_{N'N'} \end{bmatrix},$$ (4.17a)

$$B' = \det [h'_{ij}],$$ (4.17b)

$$C' = mn \det \begin{bmatrix} u'_{11} & u'_{12} & h'_{13} & \cdots & h'_{1N'} \\ u'_{21} & u'_{22} & & & \vdots \\ \vdots & \vdots & \vdots & \vdots & \vdots \\ u'_{N'1} & u'_{N'2} & h'_{N'3} & \cdots & h'_{N'N'} \end{bmatrix},$$ (4.17c)

$$D' = m \det \begin{bmatrix} u'_{11} & h'_{12} & \cdots & h'_{1N'} \\ u'_{21} & & & \vdots \\ \vdots & \vdots & \vdots & \vdots \\ u'_{N'1} & h'_{N'2} & \cdots & h'_{N'N'} \end{bmatrix},$$ (4.17d)

where

$$N' \triangleq M + 2.$$ (4.17e)

Equation (4.16) shows that A', B', C', and D' are quantities proportional to the so-called transfer parameters A, B, C, and D of the equivalent two-port circuit. To make the reciprocity condition $[(AD - BC)^{1/2} = 1]$ hold, we should divide A', B', C', and D' by $(A'D' - B'C')^{1/2}$ to obtain A, B, C, and D, respectively, as

$$\begin{bmatrix} A & B \\ C & D \end{bmatrix} = \frac{1}{\sqrt{A'D' - B'C'}} \begin{bmatrix} A' & B' \\ C' & D' \end{bmatrix}.$$ (4.18)

4.4 Short-Boundary Planar Circuit Having Two Waveguide Coupling Ports

4.4.1 Basic Equation

We next consider a waveguide-coupled circuit as shown in Fig. 4.1b, but having a more arbitrary peripheral shape. The periphery of the circuit is again divided and numbered, as shown in Fig. 4.4, as follows:

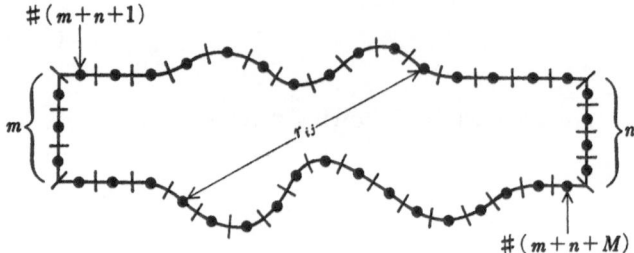

Fig. 4.4. Symbols used in the computer analysis $-$ II (case of the waveguide-coupled type)

1) Input Port 1: $i = 1 \sim m$;

2) Output Port 2: $i = (m+1) \sim (m+n)$;

3) Circuit periphery: $i = (m+n+1) \sim (m+n+M)$.

Thus $m+n+M (=N)$ sampling points are provided. The derivation of the $N \times N$ matrix equation (4.5) follows entirely the same process as was described in the preceding section.

4.4.2 Simplification of the Basic Equation

We now assume for simplicity that straight waveguide sections with appropriate length exist on both sides of the input and output planes, so that only the TE_{10} mode exists at those planes. (A comment on this assumption is given at the end of this subsection.)

Under the above assumption, we may relate $V_1 \sim V_m$ and $I_1 \sim I_m$, $V_{m+1} \sim V_{m+n}$ and $I_{m+1} \sim I_{m+n}$ with simple proportionality formulas. We use representative values of those voltages and currents defined as V_p and V_q: voltages at the center of the input and output reference planes, respectively; and I_p and I_q: currents flowing across the entire widths of the input and output reference planes, respectively. Since the variation of the voltage and current across the reference planes are both sinusoidal, we may write

$$
\left\{
\begin{aligned}
V_1 &= \alpha_1 V_p \\
V_2 &= \alpha_2 V_p \\
&\;\;\vdots \\
V_m &= \alpha_m V_p
\end{aligned}
\right.
\qquad
\left\{
\begin{aligned}
I_1 &= \alpha_1 I_p/m \\
I_2 &= \alpha_2 I_p/m \\
&\;\;\vdots \\
I_m &= \alpha_m I_p/m
\end{aligned}
\right. ,
\qquad\qquad (4.19)
$$

where

$$
\alpha_i = \sin\left(\frac{2i-1}{2m}\pi\right), \qquad \text{and}
$$

$$
\begin{cases}
V_{m+1} = \beta_1 V_q \\
V_{m+2} = \beta_2 V_q \\
\quad\vdots \\
V_{m+n} = \beta_n V_q
\end{cases}
\quad
\begin{cases}
I_{m+1} = \beta_1 I_q/n \\
I_{m+2} = \beta_2 I_q/n \\
\quad\vdots \\
I_{m+n} = \beta_n I_q/n
\end{cases},
\tag{4.20}
$$

where

$$
\beta_i = \sin\left(\frac{2i-1}{2n}\,\pi\right).
$$

Using the previous relations, we may rewrite (4.6) as

$$
[U]
\left.\begin{bmatrix}
\alpha_1 V_p \\
\vdots \\
\alpha_m V_p \\
\hline
\beta_1 V_q \\
\vdots \\
\beta_n V_q \\
\hline
0 \\
\vdots \\
0
\end{bmatrix}
\begin{array}{l}
\Big\}\,m \\[18pt]
\Big\}\,n = [H] \\[18pt]
\Big\}\,M
\end{array}\right.
\left.\begin{bmatrix}
\alpha_1 I_p/m \\
\vdots \\
\alpha_m I_p/m \\
\hline
\beta_1 I_q/n \\
\vdots \\
\beta_n I_q/n \\
\hline
I_{m+n+1} \\
\vdots \\
I_{m+n+M}
\end{bmatrix}
\begin{array}{l}
\Big\}\,m \\[18pt]
\Big\}\,n \\[18pt]
\Big\}\,M
\end{array}\right..
\tag{4.21}
$$

Following the procedure described in the preceding section, we again define the following $(M+2)\times(M+2)$ matrices to decrease the number of constraints, as

$$
[U'] =
\left[
\begin{array}{ccc|c}
\displaystyle\sum_{i=1}^{m}\sum_{j=1}^{m}\alpha_j u_{ij} &
\displaystyle\sum_{i=1}^{m}\sum_{j=m+1}^{m+n}\beta_{j-m}u_{ij} &
\displaystyle\sum_{i=1}^{m}u_{i(m+n+1)}\cdots \\[20pt]
\displaystyle\sum_{i=m+1}^{m+n}\sum_{j=1}^{m}\alpha_j u_{ij} &
\displaystyle\sum_{i=m+1}^{m+n}\sum_{j=m+1}^{m+n}\beta_{j-m}u_{ij} &
\displaystyle\sum_{i=m+1}^{m+n}u_{i(m+n+1)}\cdots \\[20pt]
\hline
\displaystyle\sum_{j=1}^{m}\alpha_j u_{(m+n+1)j} &
\displaystyle\sum_{j=m+1}^{m+n}\beta_{j-m}u_{(m+n+1)j} &
u_{\substack{m+n+1\\m+n+1}}\cdots \\[10pt]
\vdots & \vdots & \vdots \\
& & \cdots\cdots u_{\substack{m+n+M\\m+n+M}}
\end{array}
\right]
\tag{4.22}
$$

$$
[H'] = \text{(similar to the above)} .
\tag{4.23}
$$

Using the previous matrices, we may rewrite (4.21) in a form identical to (4.9). The transfer parameters are also derived in the same way.

A comment is added here as to the assumption made at the beginning of this subsection, i.e., only the TE_{01} mode exists at input/output planes because straight waveguide sections with appropriate length are assumed to exist on both sides of these planes. We may remove half of this assumption, i.e., the presence of the straight waveguide sections "inside" the reference planes, if we consider $(m-1)$ and $(n-1)$ higher-order modes in the input and output waveguides, respectively. Then we may take into account the reactive waveguide impedances for those higher-order modes, and no longer need the reduction of the constraints performed to obtain (4.22, 23). Such an analysis of short-boundary planar circuits will be described in Sect. 4.6.

4.5 Examples of Numerical Analysis

To show the validity and accuracy of the methods of analysis presented so far, two examples of numerical analysis whose results can be compared with analytical solutions will first be described. Next, several practical circuits are analyzed, and the results are compared with experimental data or other theories reported in the literature.

4.5.1 Short-Circuited Radial Line

The first example is of the coaxial-coupled type: a thin radial line terminated by a short-circuit wall (Fig. 4.5). For such a one-port circuit, the driving point impedance is given, instead of by (4.16), by

$$-Z_p = B'/D' . \tag{4.24}$$

This driving point impedance has been computed for dimensions (Fig. 4.5) $R_i = 1$ mm, $R_0 = 10$ mm, and $d = 0.5$ mm, at 21 frequencies ranging between $k\,(= 2\pi f/c) = 200 \sim 400$ m^{-1}. Various numbers of divisions ($M = 10, 20, 30$ for $m = 5$, and $M = 20, 30, 40$ for $m = 10$) have been used to obtain the estimate of the computation error.

The results of the cases for $m = 10$ are shown in Fig. 4.6. The abscissa and the ordinate show the real and imaginary parts of the driving point impedance, respectively. Note, however, that the abscissa is expanded by a factor of 100 to exaggerate the computation error. Small circles upon the ordinate show the theoretical values [4.6]

$$-Z_p = -j\frac{Z_{0i}d}{2\pi R_i}\,\frac{\sin(\theta_i-\theta_o)}{\cos(\phi_i-\theta_o)} , \tag{4.25}$$

where subscripts i and o represent the inner and outer boundaries, respectively; θ and ϕ are the quantities defined by

$$H_0^{(1)}(x) = J_0(x) + jN_0(x) = G_0(x)\exp[j\theta(x)] ,$$

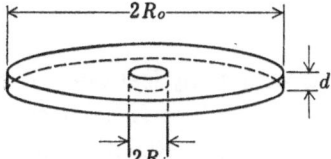

Fig. 4.5. A radial line

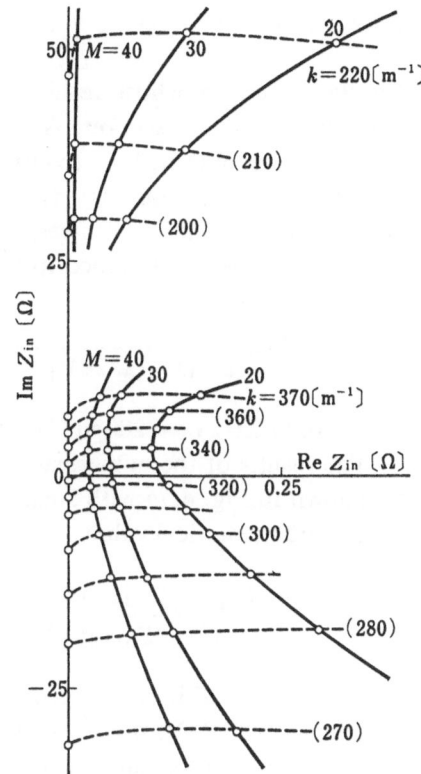

Fig. 4.6. The computed input impedance of a
radial line. Note that the real part (abscissa) is
exaggerated to show the computational error
[4.1]

$$jH_1^{(1)}(x) = -N_1(x) + jJ_1(x) = G_1(x)\exp[j\phi(x)] ,$$

where $x = kr$; and Z_{0i} denotes the wave impedance at the inner boundary, defined as

$$Z_{0i} = 120\pi[G_0(kR_i)/G_1(kR_i)] \quad [\Omega] .$$

The bracketed numerals in Fig. 4.6 denote the wavenumber in m^{-1}. Some of the broken curves ($k = 200, 210, 220\ m^{-1}$) seem to converge for increasing M to values somewhat different from theoretical ones. This is natural because m (the sampling-point number around the port) is finite.

The estimate of the error obtained from Fig. 4.6 is several tenths Ω for the real part, and several Ω (in most cases below 2 Ω) for the imaginary part. The series resonance frequency obtained with the numerical analysis is 15.68 GHz ($k = 328\ m^{-1}$) for $m = 10$, $M = 40$, which shows an error of approximately 1% as compared with the theoretical value 15.87 GHz ($k = 332\ m^{-1}$).

4.5.2 Uniform Waveguide Section

Another example, whose results can be compared with analytical ones, is a uniform waveguide section. A finite section of the standard X-band rectangular waveguide with a (width) = 22.9 mm, b (height) = 10.2 mm, and l (length) = $2a$ = 45.8 mm is assumed to be terminated by a resistive sheet placed perpendicularly with respect to the waveguide axis; the surface resistance of the sheet is assumed to be equal to the waveguide impedance

$$Z_0 = 120\,\pi\,\frac{b}{a}\,\frac{1}{[1-(\lambda/2a)^2]^{1/2}} \quad [\Omega] \tag{4.26}$$

at $f = 10.00$ GHz ($k = 209$ m^{-1}), which is 222.2 Ω.

The results of the numerical computation of the driving point impedance are shown in Fig. 4.7a with small dots, for $m = n = 10$, $M = 40$. In the same figure, the theoretical value

$$Z_{in} = Z_0\,\frac{(Z_L/Z_0)+j\tan(2\pi l/\lambda_g)}{1+j(Z_L/Z_0)\tan(2\pi l/\lambda_g)} \tag{4.27}$$

is also plotted with small crosses where Z_L, Z_0, and λ_g denote the load impedance (222.2 Ω), the waveguide impedance at the given frequency, and the wavelength in the guide, respectively. The complicated part (below, right) of Fig. 4.7a is shown in Fig. 4.7b in an enlarged scale.

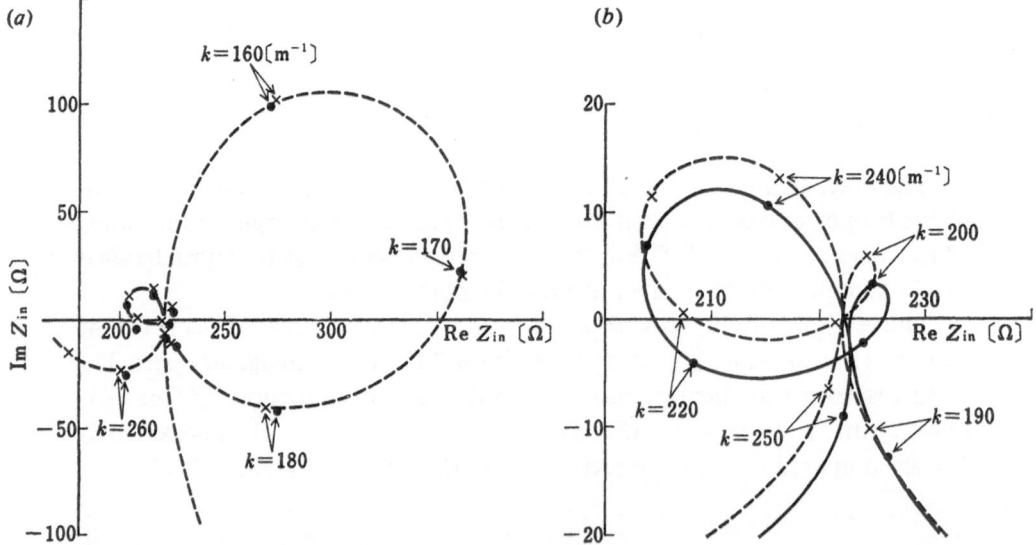

Fig. 4.7a, b. The computed input impedance of a waveguide section: (a) the overall frequency characteristics, (b) an enlarged figure of a portion [4.1]

4.5.3 Waveguide Section Including a Thick Inductive Window

In the following three subsections, more practical circuits are analyzed numerically; the obtained results are compared with the approximate analyses and experimental data. In those numerical analyses, for convenience of comparison, the obtained transfer parameters are once converted into equivalent T-circuit parameters. This conversion process will first be described.

We consider a waveguide section including a symmetrical obstacle, a thick inductive window as in the example shown in Fig. 4.8a, and represent the entire section by three equivalent circuits: those of two straight sections and an equivalent T-circuit of the obstacle itself.

If we denote the transfer parameters of the entire section, the straight sections and the obstacle by F_c, F_L, and F_T, respectively, we have

$$[F_c] = [F_L][F_T][F_L] , \quad \text{where} \tag{4.28}$$

$$[F_L] = \begin{bmatrix} \cos(2\pi L/\lambda_g) & jZ_0\sin(2\pi L/\lambda_g) \\ j(1/Z_0)\sin(2\pi L/\lambda_g) & \cos(2\pi L/\lambda_g) \end{bmatrix} . \tag{4.29}$$

From the elements of F_c computed numerically and those of F_L, the elements of F_T are obtained as

$$[F_T] \triangleq \begin{bmatrix} A_T & B_T \\ C_T & D_T \end{bmatrix} = [F_L]^{-1}[F_c][F_L]^{-1} . \tag{4.30}$$

On the other hand, F_T is expressed by the T-circuit parameters (Fig. 4.8b) as

$$[F_T] = \frac{1}{Z_a}\begin{bmatrix} Z_a+Z_b & 2Z_aZ_b+Z_b^2 \\ 1 & Z_a+Z_b \end{bmatrix} . \tag{4.31}$$

Hence, the normalized T-circuit parameters are given in terms of the computed elements of F_T as

$$\frac{Z_a}{Z_0} = \frac{1}{C_T Z_0}\left(= j\frac{X_a}{Z_0}\right) , \tag{4.32}$$

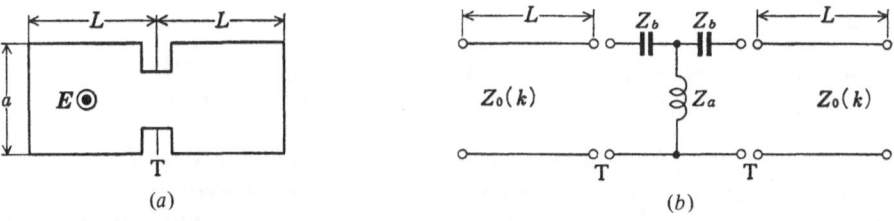

Fig. 4.8a, b. Equivalent circuit representation of a symmetrical waveguide obstacle: (a) a thick inductive window (an example), (b) the equivalent T-circuit representation

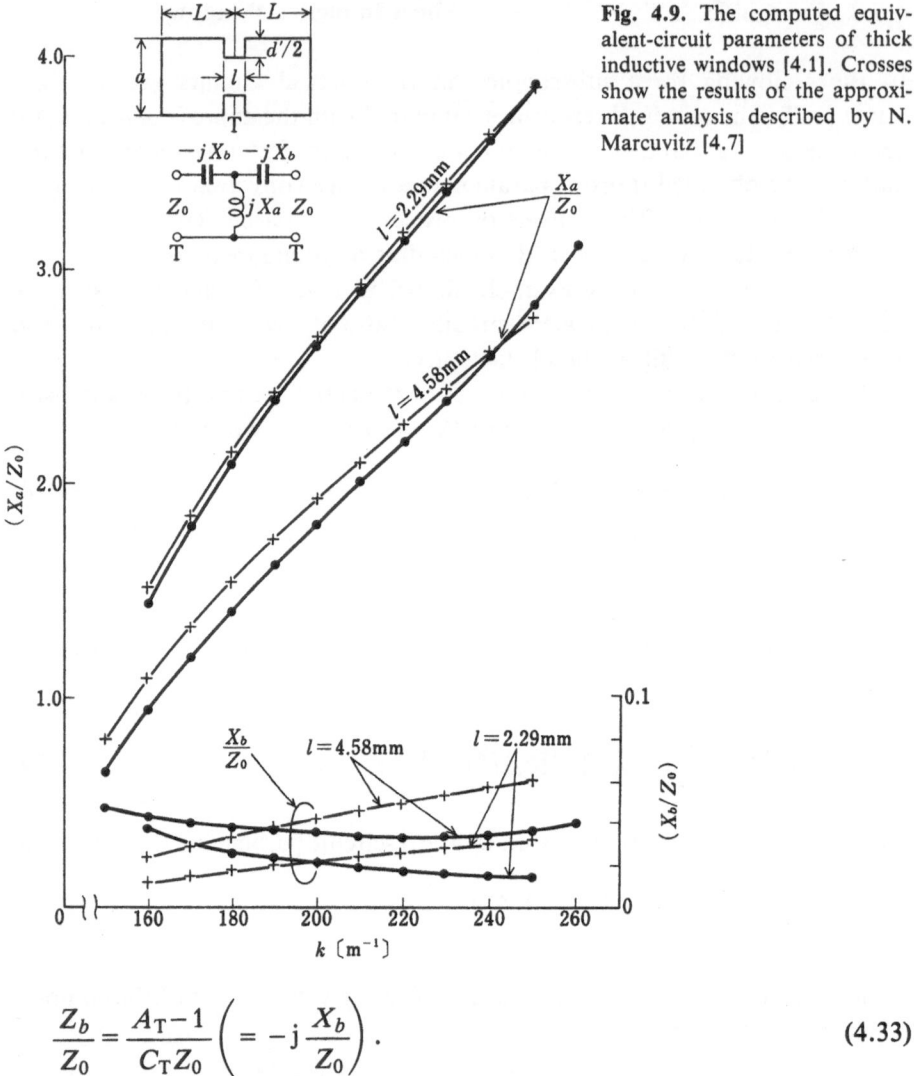

Fig. 4.9. The computed equivalent-circuit parameters of thick inductive windows [4.1]. Crosses show the results of the approximate analysis described by N. Marcuvitz [4.7]

$$\frac{Z_b}{Z_0} = \frac{A_T - 1}{C_T Z_0}\left(= -j\frac{X_b}{Z_0} \right). \tag{4.33}$$

Figure 4.9 shows the results of computer analyses of the frequency characteristics of thick inductive windows. The ordinate gives the normalized T-circuit parameters X_a/Z_0 and X_b/Z_0. The dimensions of the circuit are assumed to be

Case 1: $a = 22.9$ mm, $d'/2 = 2.29$ mm, $l = 2.29$ mm, $L = 24.045$ mm.
Case 2: $a = 22.9$ mm, $d'/2 = 2.29$ mm, $l = 4.58$ mm, $L = 25.19$ mm.

The total sampling-point number is 66 in Case 1 and 68 in Case 2.

The dots in Fig. 4.9 show the results of the computer analyses. The crosses show the approximate theoretical values described in [4.7]. The difference between dots and crosses is found to be less than 0.02. (The errors in X_b/Z_0

might seem rather large; however, note that the scales of the ordinate for X_b/Z_0 and X_a/Z_0 are different.)

4.5.4 Waveguide Sections Including Corners

The dots in Fig. 4.10 show the results of the computer analyses of 30° and 60° waveguide corners. The ordinate again shows X_a/Z_0 and X_b/Z_0. The circuit dimensions are $a = 22.9$ mm and $L = 22.9$ mm, and the total sampling-point number $N = 66$ for $\alpha = 30°$ and $N = 72$ for $\alpha = 60°$. Small crosses in the figure show the experimental data found in [4.7].

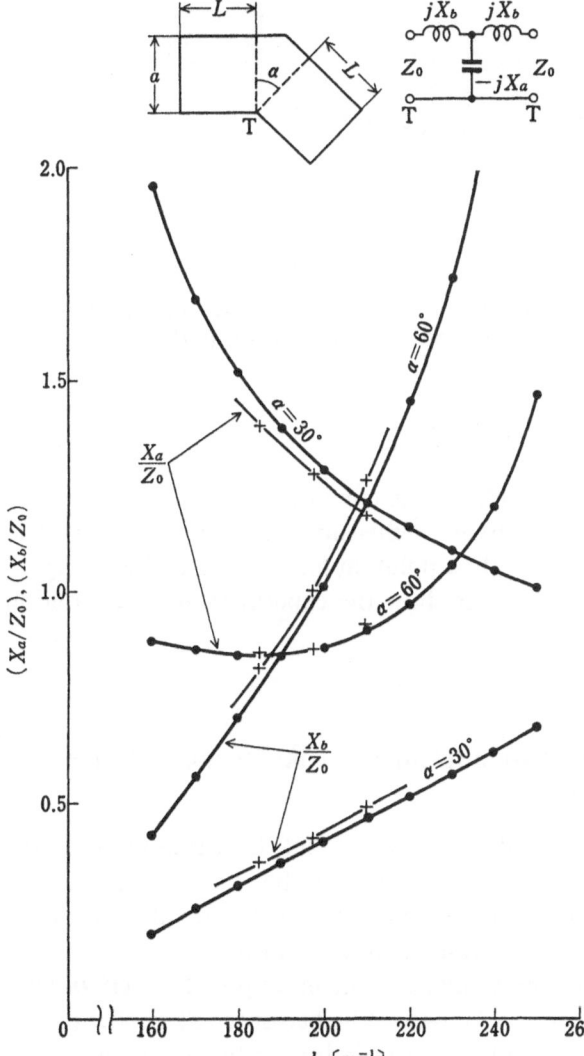

Fig. 4.10. The computed equivalent-circuit parameters of waveguide corners [4.1]. Crosses show the experimental data described by N. Marcuvitz [4.7]

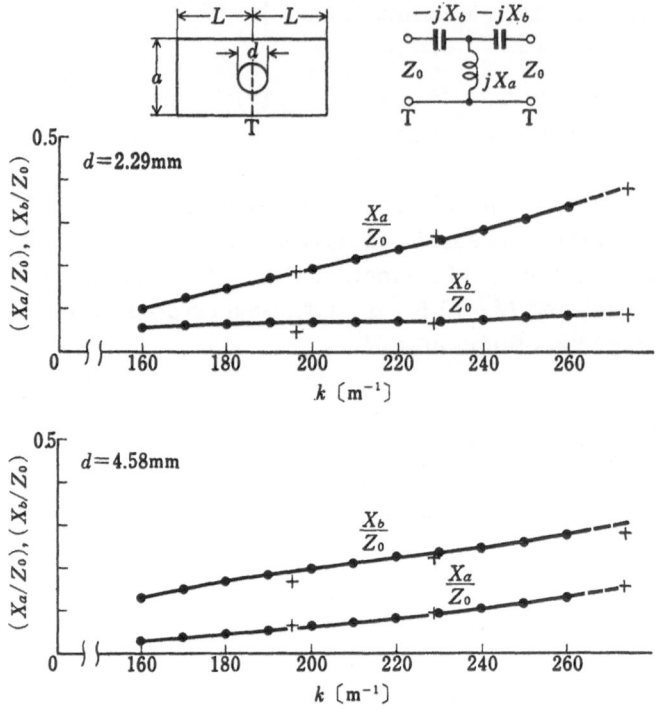

Fig. 4.11. The computed equivalent-circuit parameters of a post in a waveguide [4.1]. Crosses show the results of the approximate analysis described by N. Marcuvitz [4.7]

4.5.5 Waveguide Section Including Post

The dots in Fig. 4.11 show the results of the computer analyses of waveguide sections including post. The circuit dimensions are $a = 22.9$ mm and $L = 22.9$ mm, and the total sampling-point number $N = 72$, including 12 points around the post. Small crosses show the experimental data found in [4.7].

4.6 Higher-Order Mode Consideration at Reference Planes

In the final paragraph of Sect. 4.4.2, it is stated that the assumption of "the presence of the straight waveguide sections 'inside' the reference planes" can be removed by considering higher-order modes at the reference planes. A brief description of this computational technique [4.8] follows.

We consider that at the reference plane of the input port 1, where m sampling points are provided (Sect. 4.4.1), $(m-1)$ higher-order TE modes (TE_{20}, TE_{30}, ..., TE_{m0} modes) exist in addition to the fundamental TE_{10} mode. In the external waveguides, the TE_{10} mode can be propagated, whereas TE_{j0} modes ($j = 2, 3, \ldots, m$) can not be propagated. Hence, the latter modes are said to be terminated by reactive characteristic impedances. In the following, to avoid confusion in symbols, $(-1)^{1/2}$ is denoted as (j).

We set the x coordinate along the reference plane of the port 1 and assume that the width of the port is a. We denote the spatial peak voltage and spatial peak current density of the jth mode by V_{pj} and i_{pj}, respectively. Then the voltage and current density of the jth mode at x are expressed as

$$V_j(x) = V_{pj}\sin\left(\frac{j\pi x}{a}\right),$$ (4.34)

$$i_j(x) = i_{pj}\sin\left(\frac{j\pi x}{a}\right).$$ (4.35)

If we further define the voltage V_j and current I_j of the TE_{j0} mode as

$$V_j = \langle |V_j(x)| \rangle = \frac{1}{(a/j)}\int_0^{(a/j)} V_j(x)\,dx,$$ (4.36)

$$I_j = \int_0^{(a/j)} i_j(x)\,dx,$$ (4.37)

where $\langle\,\rangle$ denotes an average across the reference plane, then the characteristic impedance of the TE_{j0} mode is given as [4.9]

$$Z_{j0} = \frac{V_j}{I_j} = (j)\,120\,\pi\,\frac{d}{(a/j)}\,\frac{1}{\sqrt{(j\lambda/2a)^2 - 1}}.$$ (4.38)

On the other hand, the voltage and current at the ith sampling point ($i = 1, 2, \ldots, m$) contributed by the jth mode are expressed, respectively, as

$$V_{ij} = \alpha_{ij}V_j$$ (4.39a)

$$\left[\alpha_{ij} = \frac{m}{j}\sin\left(\frac{2i-1}{2m}j\pi\right)\sin\left(\frac{j\pi}{2m}\right)\right],$$ (4.39b)

$$I_{ij} = \beta_{ij}I_j$$ (4.40a)

$$\left[\beta_{ij} = \sin\left(\frac{2i-1}{2m}j\pi\right)\sin\left(\frac{j\pi}{2m}\right)\right].$$ (4.40b)

Therefore, the voltage and current at the ith sampling point, V_i and I_i, are given as

$$[V_i] = [\alpha_{ij}][V_j],$$ (4.41)

$$[I_i] = [\beta_{ij}][I_i] ,$$ (4.42)

where symbol [ ] denotes a column vector.

On the other hand, the voltage and current column vectors $[V_j]$ and $[I_j]$ are related as

$$
\begin{bmatrix} V_1 \\ \hline V_2 \\ \vdots \\ V_m \end{bmatrix}
=
\left[\begin{array}{c|ccccc}
1 & & & 0 & & \\ \hline
 & -Z_{20} & & & & \\
0 & & -Z_{30} & & 0 & \\
 & & 0 & \ddots & & \\
 & & & & & -Z_{m0}
\end{array} \right]
\begin{bmatrix} V_1 \\ I_2 \\ I_3 \\ \vdots \\ I_m \end{bmatrix} .
$$ (4.43)

Substituting (4.41 – 43) into the upper part of (4.21) [in this case, into the upper $(m+n)$ elements of the column vectors in (4.21)], and rearranging the matrix elements, we may derive the transfer parameters in a manner similar to, but somewhat more complicated than, that in Sect. 4.4 [4.8].

4.7 Summary

The basic formulation for the computer analysis of arbitrarily shaped, short-boundary planar circuits has been presented. The validity and error of the proposed method of analysis have been shown through comparison with other theories and experiments. The required computer time is not yet short; for example, more than one minute is required to obtain all the dots in Fig. 4.9 using a typical high-speed computer. The improvement of the program toward shorter computer time and better accuracy is left for further efforts.

5. Segmentation Method

In many practical planar circuitries, the circuit pattern can be divided into several segments which themselves have simpler shapes such as squares, rectangles, triangles, circular sectors, or annular sectors. In such cases, an efficient analytical approach different from both the Green's-function approach and the contour-integral approach can be used. This approach, in which the characteristics of a planar circuit are computed by combining those of the segmented elements, is called the segmentation method. It features a relatively short required computer time. In this chapter, the principle and computer algorithm of the segmentation method are described. As seen in the next chapter, this method is also useful in the trial-and-error optimum design of a planar circuit because of the short computer time required.

5.1 Background

Several methods have been presented in the preceding chapters for the analysis of planar circuits. When the circuit pattern is simple — i.e., square, rectangular, triangular, circular, or annular — the impedance matrix can be obtained in a series-expansion form from the Green's function of the wave equation (Chap. 2). When the circuit pattern is more arbitrary, the numerical analysis based upon the contour-integral representation of the wave equation is most efficient (Chaps. 3, 4). Other numerical approaches applicable to arbitrary circuit patterns are the variational method (Sects. 3.7.4, 5), relaxation method, and the finite-element method (Sect. 3.7.6).

When the circuit pattern is entirely arbitrary, we must rely upon one of those numerical analyses. However, in actual planar circuitry, an entirely arbitrary circuit pattern is not very common; in many cases the pattern consists of several "segments" which themselves have simpler shapes such as squares, rectangles, triangles, circular sectors, or annular sectors.

The *segmentation method* described in this chapter is a method in which the characteristics of a planar circuit are computed by combining those of the segmented elements. It features relatively short computer time required in the analysis.

In principle, the segmentation method can be applied to both a triplate-type, open-boundary planar circuit and a waveguide-type, short-boundary planar circuit. However, in the following description, only the triplate-type circuit is considered.

5.2 Theory of Segmentation Method Using S Matrices

5.2.1 Basic Concepts

The basic equation governing the electromagnetic field in a triplate-type planar circuit is the two-dimensional wave equation

$$(\nabla_T^2 + k^2) V = 0, \quad \text{where} \tag{5.1a}$$

$$\nabla_T^2 = \frac{\partial^2}{\partial x^2} + \frac{\partial^2}{\partial y^2}, \quad V = E_z d, \quad k^2 = \omega^2 \varepsilon \mu; \tag{5.1b}$$

ε and μ denote the permittivity and permeability of the spacing material, ω the angular frequency, and d the spacing between conductors (Sects. 2.2.1, 2).

To analyze such a circuit means to solve the wave equation under given boundary conditions, and determine the circuit parameters of its equivalent n-port network as functions of the frequency. When the circuit pattern is simple (for instance, rectangular), the problem may be solved analytically (Chap. 2). The segmentation method is a method of analysis suited to somewhat more complex patterns, as shown in Fig. 5.1a. We divide such a circuit pattern into simpler "segments" (Fig. 5.1b), compute the characteristics of each rect-

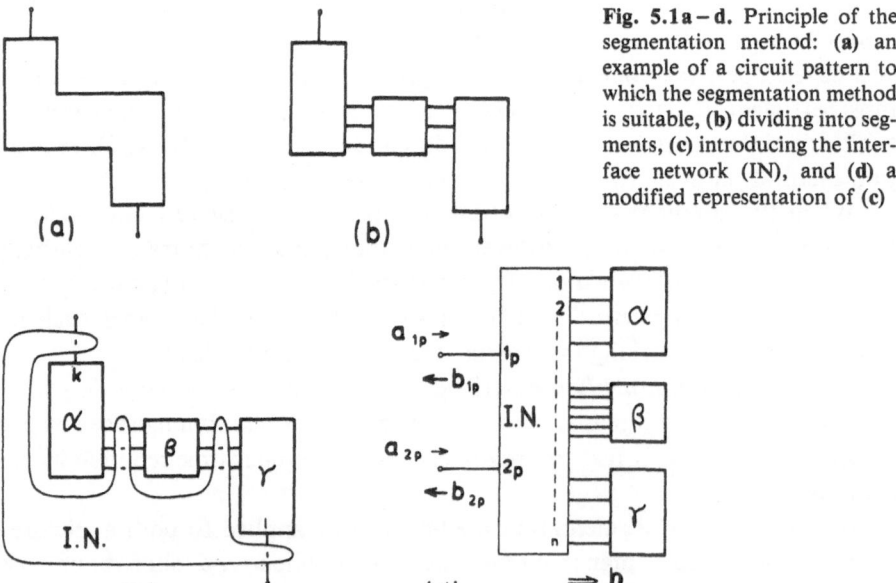

Fig. 5.1a–d. Principle of the segmentation method: (a) an example of a circuit pattern to which the segmentation method is suitable, (b) dividing into segments, (c) introducing the interface network (IN), and (d) a modified representation of (c)

angular segment from its Green's function (Sects. 2.3, 4), and finally obtain the characteristics of the entire circuit by synthesizing the characteristics of the segments.

Several different formulations of the segmentation-method analysis have been proposed [5.1 – 4]. In this section, those using S matrices are described first: one proposed by *Okoshi* et al. [5.3] and the other (computer-time-saving scheme) by *Chadha* and *Gupta* [5.4]. Finally, an alternative approach using Z matrices is presented [5.4].

5.2.2 Interface Network

To facilitate the synthesis of the segments in a unified manner, a new circuit called the *interface network* is introduced. As shown in Fig. 5.1c, all connections between segments are made through this interface network; it only connects directly those couples of ports facing each other. However, a unified formulation of the synthesis is only possible with the aid of this fictitious network.

In the following, we consider a two-port circuit for the simplicity of the formulation. Figure 5.1c may also be expressed as Fig. 5.1d. We define, with respect to Ports $1p$, $2p$, 1, 2, ..., n, incident and reflected waves of the interface network as

$$a_{1p}, a_{2p}, a_1, a_2, \ldots, a_n \quad \text{and}$$

$$b_{1p}, a_{2p}, b_1, b_2, \ldots, b_n.$$

In numbering ports, we use (numerals $+ p$) for those of the interface network connected to external ports, and simple numerals for those connected to segmentary circuits.

The aforementioned incident and reflected waves are related in terms of the scattering matrix S_N of the interface network as

$$
\begin{bmatrix} b_{1p} \\ b_{2p} \\ b_1 \\ b_2 \\ \vdots \\ b_n \end{bmatrix} = [S_N] \begin{bmatrix} a_{1p} \\ a_{2p} \\ a_1 \\ a_2 \\ \vdots \\ a_n \end{bmatrix},
\tag{5.2}
$$

where elements of $[S_N]$ are given by

$$
[S_N]_{ij} = \begin{cases} 1, & \text{if Ports } i \text{ and } j \text{ are connected} \\ 0, & \text{if Ports } l \text{ and } j \text{ are not connected}, \end{cases}
\tag{5.3}
$$

where $i, j = 1p$, $2p$, 1, 2, ..., n. When $i = j$, $[S_N]_{ij} = 0$.

5.2.3 Computation of the S Matrix

If Ports $1p$ and $2p$ are connected directly in the interface network to Ports k and l, respectively, S_N may be expressed as

$$[S_N] = \begin{bmatrix} 0 & 0 & S_1^{\rightarrow} \\ 0 & 0 & S_2^{\rightarrow} \\ \hline S_1^{\downarrow} & S_2^{\downarrow} & S_n \end{bmatrix}, \tag{5.4a}$$

where $S_1^{\rightarrow}$ and $S_2^{\rightarrow}$ are row vectors, and $S_1^{\downarrow}$ and $S_2^{\downarrow}$ are column vectors defined by

$$S_1^{\rightarrow} = (0, \ldots, 0, \underset{k}{1}, 0, \ldots, 0), \quad S_2^{\rightarrow} = (0, \ldots, 0, \underset{l}{1}, 0, \ldots, 0), \tag{5.4b}$$

$$S_1^{\downarrow} = [S_1^{\rightarrow}]^t, \quad S_2^{\downarrow} = [S_2^{\rightarrow}]^t, \tag{5.4c}$$

where the superscript t denotes transposition. The submatrix S_n in (5.4a) expresses connections between segments.

The S matrix of the entire circuit should be a 2×2 matrix giving b_{1p} and b_{2p} in terms of only a_{1p} and a_{2p}. However, we should note that, because $1p$ and $2p$ are connected to k and l, respectively,

$$b_{1p} = a_k, \quad b_{2p} = a_l \tag{5.5}$$

hold. Therefore, in the following we compute a_k and a_l in terms of a_{1p} and a_{2p}.

From (5.2), taking out n rows from the bottom, we obtain

$$\begin{bmatrix} b_1 \\ b_2 \\ \vdots \\ b_n \end{bmatrix} = [S_n] \begin{bmatrix} a_1 \\ a_2 \\ \vdots \\ a_n \end{bmatrix} + S_1^{\downarrow} a_{1p} + S_2^{\downarrow} a_{2p}. \tag{5.7}$$

On the other hand, we may also write

$$\begin{bmatrix} a_1 \\ a_2 \\ \vdots \\ a_n \end{bmatrix} = [S_c] \begin{bmatrix} b_1 \\ b_2 \\ \vdots \\ b_n \end{bmatrix}, \tag{5.6}$$

where S_c is the composite S matrix of the segmentary circuits expressed as

$$[S_c] = \begin{bmatrix} S_\alpha & & 0 \\ & S_\beta & \\ 0 & & S_\gamma \end{bmatrix}, \tag{5.8}$$

S_α, S_β, and S_γ denoting the S matrix of the segments. When the segments have simple shapes (e.g., rectangular), these matrices can be obtained analytically (Chap. 2).

Eliminating the b vector from (5.6, 7), we obtain

$$\begin{bmatrix} a_1 \\ a_2 \\ \vdots \\ a_n \end{bmatrix} = TS_1^{\downarrow} a_{1p} + TS_2^{\downarrow} a_{2p}, \tag{5.9a}$$

where

$$T = [E - S_c S_n]^{-1} [S_c], \tag{5.9b}$$

E denoting an $n \times n$ unit matrix. From (5.4b, c),

$$\begin{aligned} TS_1^{\downarrow} &= \text{the } k\text{th column of } T \\ TS_2^{\downarrow} &= \text{the } l\text{th column of } T. \end{aligned} \tag{5.10}$$

Substituting (5.5) into (5.9a), we finally obtain

$$\begin{aligned} b_{1p} &= a_k = T_{kk} a_{1p} + T_{kl} a_{2p} \\ b_{2p} &= a_l = T_{lk} a_{1p} + T_{ll} a_{2p}. \end{aligned} \tag{5.11}$$

Equation (5.11) shows that the S matrix of the entire circuit is a 2×2 sub-matrix of T; it consists of four elements of T at crosspoints between the kth, lth columns, and the kth, lth rows. Thus the circuit characteristics may be derived from S_n and S_c.

In the aforementioned derivation, the entire circuit has been assumed to be a two-port circuit. The characteristics of a multiport circuit may be obtained in a similar manner.

5.2.4 Reduction of Computer Time in S Matrix Computation

Chadha and *Gupta* [5.4] proposed a modification of the above method to reduce the computer time required, in some parts of the computation to one-half. They started from an expression a little different from (5.2, 7), that is,

$$\begin{bmatrix} b_p \\ b_c \end{bmatrix} = \begin{bmatrix} S_{pp} & S_{pc} \\ S_{cp} & S_{cc} \end{bmatrix} \begin{bmatrix} a_p \\ a_c \end{bmatrix}, \tag{5.12}$$

where a_p, b_p and a_c, b_c are column vectors consisting of the normalized wave variables at p externally and c internally connected ports. In this formulation cited from [5.5], p (for example, 2) external ports are connected directly to one of the segment circuits, without going through the interface network. The interface network interconnects only internally connected ports, so that

$$b_c = S_n a_c, \tag{5.13}$$

where S_n is the connection matrix.

It is not difficult to derive an expression for the external (for example, 2×2) S matrix for the entire circuit. We compute first the lower half of (5.12) and substitute (5.13) into it to eliminate $\boldsymbol{b}_c$. Then, the obtained $\boldsymbol{a}_p$-versus-$\boldsymbol{a}_c$ relation is substitute into the upper half of (5.12). The result is given as

$$\boldsymbol{b}_p = S_p \boldsymbol{a}_p, \quad \text{where} \tag{5.14a}$$

$$S_p = S_{pp} + S_{pc}(S_n - S_{cc})^{-1} S_{cp}. \tag{5.14b}$$

Obviously, the computation of (5.14) requires the inversion of a matrix of an order equal to the number of interconnected ports. Here we consider an example shown in Fig. 5.2 to illustrate the total computational effort needed in the segmentation method. This network is a planar circuit version of a compensated in-line power divider [5.4, 6].

In this segmentation-method computation, each connection is divided into six subports for obtaining the Z matrix. At the three external ports, six subports are combined together using ideal six-way power dividers (not shown in the figure). This computational technique [5.1, 2] is described in Appendix A5.1. To obtain S matrices of individual segments in such a planar circuit, five complex matrices, three of order 12, and one each of order 14 and 20, have to be inverted. The number of interconnections in the network is 44; hence a complex matrix of order 88 has to be inverted to obtain the overall scattering matrix.

However, as described in [5.7], if we change suitably the ordering of rows and columns, we may write the connection matrix in the form

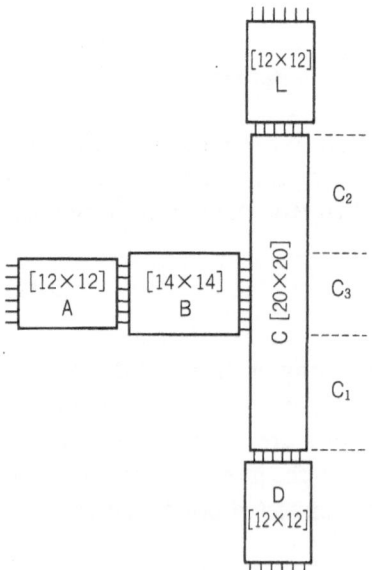

Fig. 5.2. An example of a planar network for explaining the computer-time-reduction technique [5.4]

$$S_n = \begin{bmatrix} 0 & E \\ E & 0 \end{bmatrix}, \tag{5.15}$$

where E is a unity matrix and 0 a null matrix, both of order $c/2$. It is also pointed out in [5.7] that when two segments are being interconnected, submatrix S_{cc} in (5.12) can be written in a block-diagonal form as

$$S_{cc} = \begin{bmatrix} M & 0 \\ 0 & N \end{bmatrix}, \tag{5.16}$$

where M and N are again $c/2 \times c/2$ matrices. In such cases $(S_n - S_{cc})$ in (5.14) becomes of the form

$$(S_n - S_{cc}) = \begin{bmatrix} -M & E \\ E & -N \end{bmatrix}. \tag{5.17}$$

Chadha and *Gupta* pointed out that the inversion of $(S_n - S_{cc})$ in the above form requires inversion of two matrices of order $c/2$, resulting in a saving of computational effort.

This technique can also be extended for interconnection of more than two segments [5.4]. For example, it is possible to write S_{cc} in block-diagonal form for the circuit shown in Fig. 5.2, and the computation of the S matrix now requires inversion of two complex matrices of order 44 each.

5.3 Theory of Segmentation Method Using Z Matrices

Chadha and *Gupta* also pointed out that the computational efficiency of the segmentation method can be improved if the Z matrices of individual segments, instead of the S matrices, are combined to give the overall Z matrix.

5.3.1 Basic Equations

We note first that for a lossless network, the Z matrix is purely imaginary. Multiplication and inversion of purely imaginary matrices can be performed with a computer time the same as required for real matrices.

The Z matrix of a segmented planar circuit can easily be given, corresponding to the S matrix formulation of (5.12), as

$$\begin{bmatrix} V_p \\ V_c \end{bmatrix} = \begin{bmatrix} Z_{pp} & Z_{pc} \\ Z_{cp} & Z_{cc} \end{bmatrix} \begin{bmatrix} I_p \\ I_c \end{bmatrix}, \tag{5.18}$$

where V_p, I_p and V_c, I_c are the voltages and currents at the p externally and c internally connected ports. However, in this case the function of the interface network is expressed in a form fairly different from the S matrix formulation.

The interconnection constraints are as follows: the voltages at two connected ports are equal, and the sum of currents at the two connected ports is zero. These conditions are expressed as

$$\Gamma_1 V_c = 0 , \tag{5.19a}$$

$$\Gamma_2 I_c = 0 , \tag{5.19b}$$

where Γ_1 and Γ_2 are two matrices with $c/2$ rows and c columns describing the connections. For example, if Ports 1 and 5, 2 and 6, 3 and 7, 4 and 8 are connected directly to each other,

$$\Gamma_1 = \begin{bmatrix} 1 & 0 & 0 & 0 & -1 & 0 & 0 & 0 \\ 0 & 1 & 0 & 0 & 0 & -1 & 0 & 0 \\ 0 & 0 & 1 & 0 & 0 & 0 & -1 & 0 \\ 0 & 0 & 0 & 1 & 0 & 0 & 0 & -1 \end{bmatrix} = [E \,\vdots\, -E] , \tag{5.20a}$$

$$\Gamma_2 = \begin{bmatrix} 1 & 0 & 0 & 0 & 1 & 0 & 0 & 0 \\ 0 & 1 & 0 & 0 & 0 & 1 & 0 & 0 \\ 0 & 0 & 1 & 0 & 0 & 0 & 1 & 0 \\ 0 & 0 & 0 & 1 & 0 & 0 & 0 & 1 \end{bmatrix} = [E \,\vdots\, E] . \tag{5.20b}$$

We may derive first a relation between I_c and I_p from the lower half of (5.18) and (5.19a); i.e.,

$$0 = \Gamma_1 (Z_{cp} I_p + Z_{cc} I_c) . \tag{5.21}$$

Combining this equation further with (5.19b), we have

$$\begin{bmatrix} \Gamma_1 Z_{cc} \\ j\Gamma_2 \end{bmatrix} I_c = \begin{bmatrix} -\Gamma_1 Z_{cp} \\ 0 \end{bmatrix} I_p , \tag{5.22}$$

where 0 is a $(c/2) \times p$ null matrix. The left-hand side of (5.19b) has been multiplied by j to give the lower half of (5.22), so that the matrix on the left-hand side of (5.22) becomes purely imaginary for a lossless network.

5.3.2 Computation of the Z Matrix

The impedance matrix of the entire network Z_p, which relates V_p and I_p as $V_p = Z_p I_p$, can easily be computed. We rewrite I_c in (5.18) in terms of I_p using (5.22), and finally obtain

$$Z_p = Z_{pp} - Z_{pc} \begin{bmatrix} \Gamma_1 Z_{cc} \\ j\Gamma_2 \end{bmatrix}^{-1} \begin{bmatrix} \Gamma_1 Z_{cp} \\ 0 \end{bmatrix} . \tag{5.23}$$

Obviously, this formula requires the inversion of a matrix of an order equal to the number of interconnected ports.

To analyze the circuit shown in Fig. 5.2 by the method described above, two real matrices of order 52 and 18, and a complex matrix of order 3 need be inverted. It is seen that there will be a considerable saving in the computer time as compared with the segmentation methods using S matrices. In this analysis, six subports at the three external ports are assumed to be combined together to form a single terminal pair each. The computational technique for this combination is described in Appendix A5.1.

5.3.3 Reduction of Computer Time in Z Matrix Computation

Chadha and *Gupta* showed that, by a method somewhat analogous to that described in Sect. 5.2.4, a further saving in computer time can be achieved if the connected ports are suitably subgrouped to rewrite the Z matrix formulation as follows [5.4].

The c internally connected ports are divided into groups q and r, each containing $c/2$ ports. This division must be made in such a way that q_1 and r_1 ports are connected together, q_2 and r_2 ports are connected together, and so forth. Thus the rows and/or columns for Z_{cp}, Z_{pc}, and Z_{cc} in (5.18) are reordered, and the Z matrices can be written as

$$\begin{bmatrix} V_p \\ V_q \\ V_r \end{bmatrix} = \begin{bmatrix} Z_{pp} & Z_{pq} & Z_{pr} \\ Z_{qp} & Z_{qq} & Z_{qr} \\ Z_{rp} & Z_{rq} & Z_{rr} \end{bmatrix} \begin{bmatrix} I_p \\ I_q \\ I_r \end{bmatrix} . \tag{5.24}$$

In this case the interconnections can be expressed in a much simpler form as

$$V_q = V_r , \tag{5.25a}$$

$$I_q + I_r = 0 . \tag{5.25b}$$

Substituting these equations into (5.24) and eliminating V_q, V_r, I_q, and I_r, the Z matrix of the overall network is given as

$$Z_p = Z_{pp} + (Z_{pq} - Z_{pr})(Z_{qq} - Z_{qr} - Z_{rq} + Z_{rr})^{-1}(Z_{rp} - Z_{qp}) . \tag{5.26}$$

This method using subgrouped interconnections requires the inversion of a matrix of order equal to the number of interconnections, that is, $c/2$. For example, for analyzing the circuit shown in Fig. 5.2, what we must invert are two real matrices of orders 26 and 18, and a complex matrix of order 3.

The orders of matrices that must be inverted in the computational process in various methods are compared in Table 5.1 for the circuit shown in Fig. 5.2 [5.4].

Table 5.1. Orders of matrices to be inverted in various methods of segmentation for the circuit of Fig. 5.2 [5.4]

Method	Segmentation using S matrices	Segmentation using Z matrices
Without subgrouping of the connected ports	20×20 complex $3 \times (12 \times 12)$ complex 14×14 complex 88×88 complex	52×52 real 18×18 real 3×3 complex
With suitable subgrouping of the connected ports	20×20 complex $3 \times (12 \times 12)$ complex 14×14 complex $2 \times (44 \times 44)$ complex	26×26 real 18×18 real 3×3 complex

5.4 Summary

The segmentation methods in which the characteristics of a planar circuit are computed by combining those of segmented elements have been described. There are several versions of this method. Nevertheless, they are basically grouped into those using S matrices and those using Z matrices. As seen in the next chapter, this method can be used in the trial-and-error optimum design of a planar circuit because of the short computer time required.

Recently, there was proposed a new method for analyzing a planar circuit which can be converted into a simple shape by adding one or more simple-shaped segment(s). This has been named the "desegmentation method" [5.8]. This interesting approach, however, is omitted in this volume because of space limitations; readers are encouraged to refer to [5.8] when necessary.

6. Trial-and-Error Synthesis of Optimum Planar Circuit Pattern

In the preceding chapters, only the analysis of planar circuits has been considered. As pointed out in Sect. 1.2.3, however, an important significance of the planar circuit concept lies in that, if the design technique for an arbitrarily shaped planar circuit is established in the future, it will offer an exact and efficient design tool for the microwave integrated circuit (IMC).

In Chaps. 6, 7, the design (synthesis) of the optimum planar circuit pattern is discussed. In this chapter, a synthesis technique based upon trial-and-error approach is described. In the following chapter, a more automatic, computer-oriented approach will be described. In both chapters the circuit-pattern optimization of a branch-line 3-dB hybrid circuit is considered as an example, aiming at a wider bandwidth and better symmetry in the frequency characteristics.

6.1 Background

Since the planar circuit concept was proposed for rigorous analysis and design of microwave and millimeter-wave ICs, its synthesis (determination of the circuit pattern giving prescribed circuit characteristics) has been an important technical target. In as early as 1972, it was speculated that "the design of a planar circuit, based upon the high-speed computer analysis and the trial-and-error principle, will also be possible within several years" [6.1].

However, although a number of methods were presented for the analysis in relatively early stage, i.e., by the middle of the 1970s, it took some time for the synthesis theories to appear. The first trial was reported by *Grüner* in 1974 [6.2]. He dealt with the conformal-mapping synthesis of a thin waveguide section. Since a thin waveguide section can be regarded as a short-boundary planar circuit, his method will also be applicable, if appropriately modified, to open-boundary planar circuits. However, Grüner described synthesis of only poles of transmission characteristics. Synthesis of residues must also be performed to make the synthesis complete.

The first success of the synthesis of residues was reported by *Kato* et al. in 1976 [6.3]. Their paper described a fully computer-oriented iterative synthesis of an open-boundary planar circuit having an impedance matrix with prescribed poles and residues. Basically, their aim was similar to *Grüner's* [6.2]; the differences were: (i) open-boundary problems were considered, (ii) prescribed residues were realized in addition to poles, and (iii) the mapping technique had been improved. However, this approach has not since been developed for practical applications, and hence will not be described in detail.

On the other hand, another stream of more practical synthesis technique existed since 1974; *Okoshi* et al. described a trial-and-error synthesis of a branch-line 3-dB hybrid consisting of wide striplines [6.4]. This synthesis technique, in which a designer functions as an evaluation/correction circuit in the trial-and-error loop, is never of a general nature. However, in this chapter, this primitive technique will first be described to give a basis to a more advanced, computer-oriented synthesis technique to be described in the following chapter. The trial-and-error synthesis of a branch-line 3-dB hybrid will be described as an example.

6.2 Method of Synthesis

6.2.1 Principle of the Method

We first assume a "starting" planar circuit pattern. Next we compute the frequency characteristics of that circuit, and then evaluate the obtained characteristics upon the basis of the practical requirements (broadbandness, and/or sharp cutoff, and/or symmetry, etc.). Then we modify the circuit pattern so as to minimize (or maximize) an appropriately defined evaluation function. We repeat such processes of analysis, evaluation, and modification until we obtain satisfactory characteristics.

The above principle is essentially common in both the designer-assisted, trial-and-error approach (this chapter) and the fully computer-oriented approach (next chapter). The biggest difference is found in the modification process.

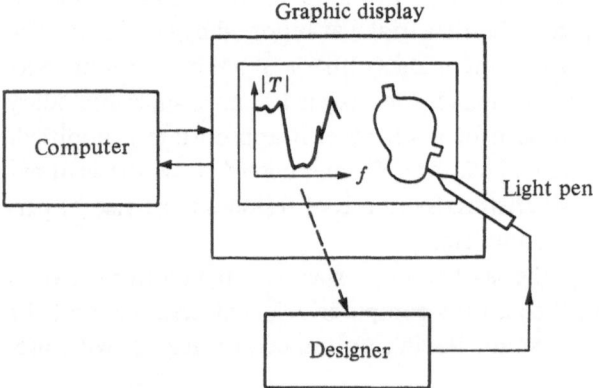

Fig. 6.1. Concept of a system for the planar circuit design

The concept of the designer-assisted approach is best illustrated in Fig. 6.1, if not realistic in the example described in the following. In the system shown in Fig. 6.1, the designer tries to modify the circuit pattern so as to optimize the characteristics. At first no "guiding principle" exists for the modification. The guiding principle itself must be discovered through the preliminary trials. A high-speed analysis technique is a premise for such a discovery. An example of such a process will be described in Sect. 6.2.5 for a 3-dB hybrid circuit. In contrast to this, in the computer-oriented approach, the modification is performed in an automatic manner.

6.2.2 Computer Analysis

We consider a branch-line 3-dB hybrid as shown in Fig. 6.2. At a frequency for which $l = \lambda_g/4$, where λ_g denotes the wavelength in the stripline, this circuit shows hybrid characteristics; i.e., when power is incident at Port 1, it will appear at Ports 2 and 3 half-and-half, while no power appears at Port 4. Similar relations hold for inputs at other ports [6.5].

However, in the following two cases, the stripline width becomes comparable to the quarter wavelength as shown in Fig. 6.3. The first case is when the required impedance level of the hybrid is relatively low as in a mixer circuit employing Schottky-barrier diodes which have low matching impedance. The

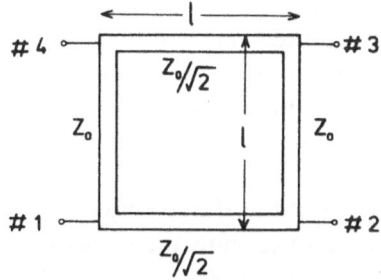

Fig. 6.2. Basic configuration of a branch-line 3-dB hybrid circuit

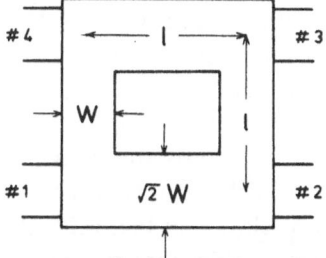

Fig. 6.3. A branch-line planar circuit which requires the planar circuit approach

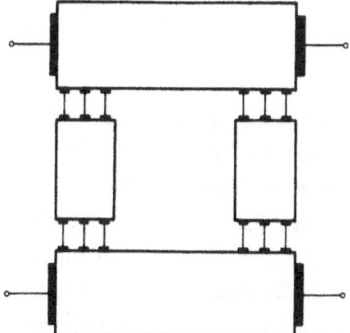

Fig. 6.4. Segmentation of the branch-line 3-dB hybrid

second case is when the frequency is relatively high and the lengths of the arms are shortened. In either case, we have to deal with the circuit as a planar circuit.

The segmentation method is most conveniently used in analyzing such a planar hybrid circuit. In the following analyses, we assume that the circuit is of triplate type, with dielectric-spacing material having the specific permittivity $\varepsilon_s = 2.53$ and the thickness $d = 1.52$ mm, in accordance with the material used in the experiment.

We first divide the circuit into four segments as shown in Fig. 6.4. Next the S matrices of each segment are computed by transforming the Z matrices given in (2.38) into S-matrix form. Finally, we synthesize them by the segmentation method described in Chap. 5 to obtain the S matrix of the entire four-port circuit.

The method of taking into account the width of external terminals (Fig. 6.4) might require some comment. When the external terminals are assumed to have certain widths, components of the impedance matrix are computed by averaging the Green's function of the wave equation over the widths of the corresponding terminals (2.31). When the terminal shifts, the averaging range should be moved accordingly.

6.2.3 Starting Circuit Pattern

As a starting circuit pattern, we take one as shown in Fig. 6.3 and assume that $l = 10.8$ mm for each arm. According to the distributed-constant model (Fig. 6.2), this length and $\varepsilon_s = 2.53$ lead to the center frequency $f_0 = 4.4$ GHz. Further, if we assume $Z_0 = 50\ \Omega$, we obtain $W = 3.6$ mm and $\sqrt{2}\,W = 5.1$ mm. Comparison of l and W suggests that the planar circuit model must be employed in exact analysis and design.

The result of the segmentation-method analysis of the aforementioned circuit pattern is shown in Fig. 6.5b. These characteristics show the increase of the center frequencies (approximately 20% above f_0) as well as asymmetries. Obviously, this starting circuit pattern requires improvement.

6.2.4 Figure of Merit

Starting at the aforementioned circuit pattern we look for better ones. As the figure of merit of a circuit pattern we use the "bandwidth" defined as follows: at any frequency within the "band", incident power at Port 1 is divided and appears at Ports 2 and 3 with percentages less than 55% and greater than 45%, and at Ports 1 and 4 with percentages less than 5%. The circuit pattern is modified successively on the trial-and-error basis so that the "bandwidth" can be maximized and the symmetry of the frequency response be optimized.

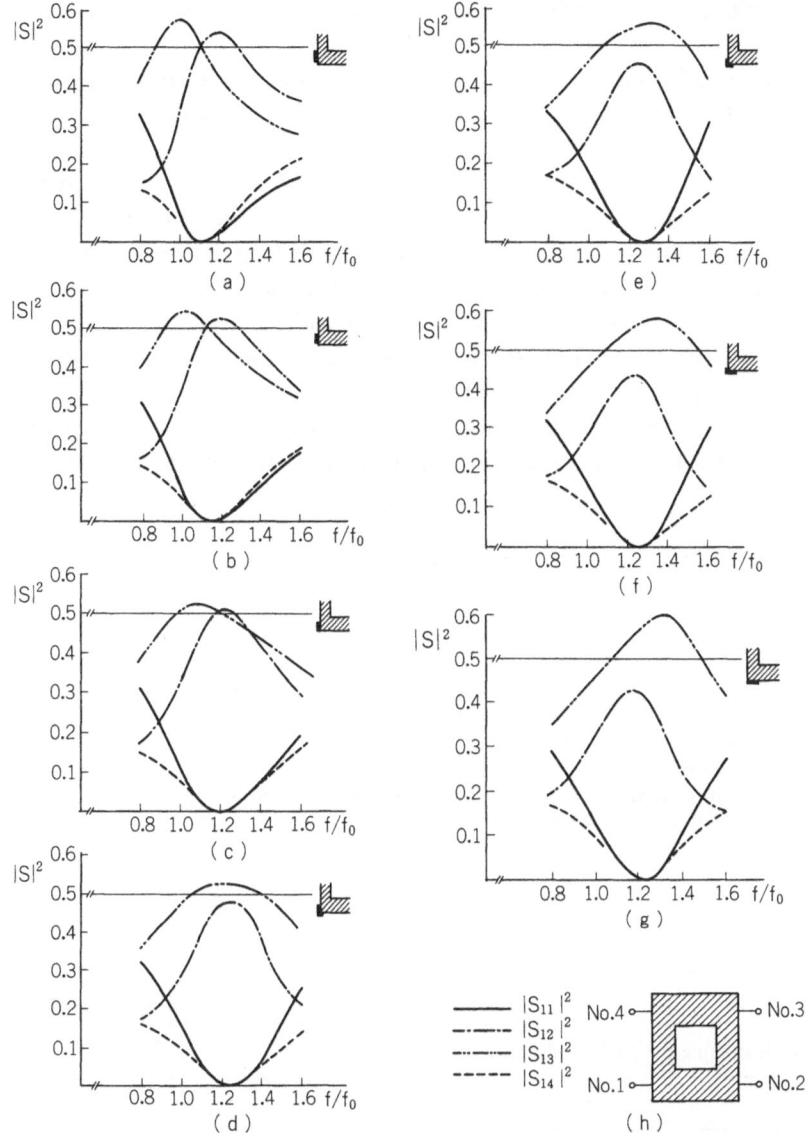

Fig. 6.5a – h. Variation of the characteristics caused by the shift of terminals: **(a – g)** the corresponding terminal positions are illustrated at the right top, **(h)** the curve designations. Symbol f_0 denotes the center frequency based upon the distributed-constant theory [6.4]

In modifying the circuit pattern, symmetries of the circuit pattern with respect to two axes remain. Hence, the circuit characteristics may be determined by four parameters: $|S_{11}|^2$, $|S_{12}|^2$, $|S_{13}|^2$, and $|S_{14}|^2$.

6.2.5 Variation of Characteristics by Modification of Circuit Pattern

To collect basic information needed in the optimization, we first investigate the variation of characteristics caused by modification of design parameters.

First, effects of the position of terminals were investigated. The computed characteristics are shown in Fig. 6.5a – g. This figure shows that when terminals move from the ends of the $\sqrt{2}\,W$-width arms towards those of the W-width arms, the S parameters vary, as shown in Table 6.1a. The optimum pattern is found somewhere between Fig. 6.5c and Fig. 6.5d. The optimum terminal position obtained is shown in Fig. 6.6, for which the center frequencies of $|S_{11}|^2$, $|S_{12}|^2$, and $|S_{14}|^2$ show the best coincidence. Hereafter, we consider only the terminal positions shown in Fig. 6.6.

Next, for further improvement of the characteristics, variations caused by shortening the $\sqrt{2}\,W$-width arms were investigated. It was found that the

Table 6.1. Variations of S parameters caused by modification of the circuit pattern [6.4]

	$|S_{11}|^2$	$|S_{12}|^2$	$|S_{13}|^2$	$|S_{14}|^2$
Center frequency (p: peak, v: valley)	↑ v	× p	↑↑ p	↑ v
Peak or valley value	–	↓	↓↑	–

(a) Effects of the shift of terminal positions. (For their shift from ends of the $\sqrt{2}\,W$-width arms towards those of the W-width arms)

	$|S_{11}|^2$	$|S_{12}|^2$	$|S_{13}|^3$	$|S_{14}|^2$
Center frequency (p: peak, v: valley)	× v	× p	↑↑ p	× v
Peak or valley value	–	×	↓↑	–

(b) Effects of shortening the $\sqrt{2}\,W$-width arms

	$|S_{11}|^2$	$|S_{12}|^2$	$|S_{13}|^2$	$|S_{14}|^2$
Center frequency (p: peak, v: valley)	× v	× p	↑ p	× v
Peak or valley value	–	↑↑	↓↓	–

(c) Effects of widening the $\sqrt{2}\,W$-width arms

↑: Increases gradually
↑↑: Increases abruptly
↓↓: Decreases abruptly
↓↑: Once decreases and then increases
×: No remarkable variation
–: These "valley" values are always almost zero

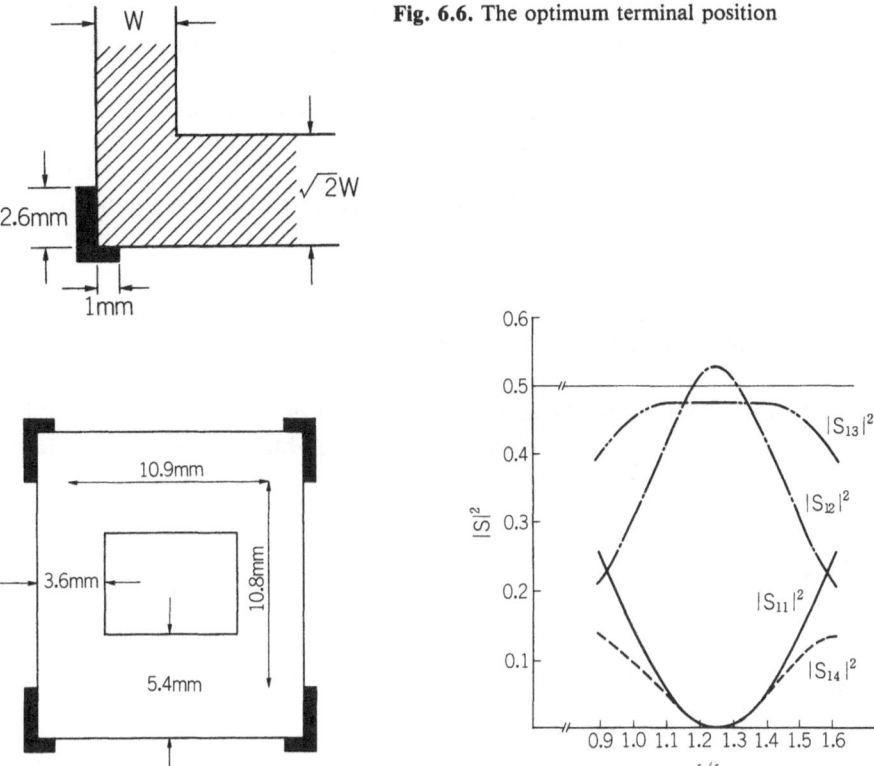

Fig. 6.6. The optimum terminal position

Fig. 6.7. The optimum circuit pattern and the corresponding characteristics [6.4]

variations as shown in Table 6.1b took place. The detailed data are omitted to save space.

Finally, the variations caused by widening the $\sqrt{2}\,W$-width arms were investigated. It was found that the variations, as shown in Table 6.1c, took place.

If we disregard a common shift of the center frequencies, the variations of the length and/or width of the W-width arms may be replaced by those of the $\sqrt{2}\,W$-width arms.

6.2.6 Optimum Circuit Pattern

Based upon the aforementioned tendencies obtained by the high-speed computer analyses using the segmentation method, we performed the final optimization as follows:

1) The optimum position of ports was determined so that center frequencies of $|S_{11}|^2$, $|S_{12}|^2$, and $|S_{14}|^2$ coincide with each other (Fig. 6.6).

2) The optimum width of the $\sqrt{2}\,W$-width arms was determined so that the peak value of $|S_{12}|^2$ is optimized. At this stage, the center frequency of $|S_{13}|^2$ is somewhat deviated.
3) The length of the $\sqrt{2}\,W$-width arms was determined so that the center frequency of $|S_{13}|^2$ coincides with the others.

The optimum circuit pattern finally obtained and the corresponding circuit characteristics are shown in Fig. 6.7.

6.3 Comparison with Experiment

To confirm the validity of the analysis and design procedures, experiment has been performed using the optimum circuit pattern obtained theoretically [6.4].

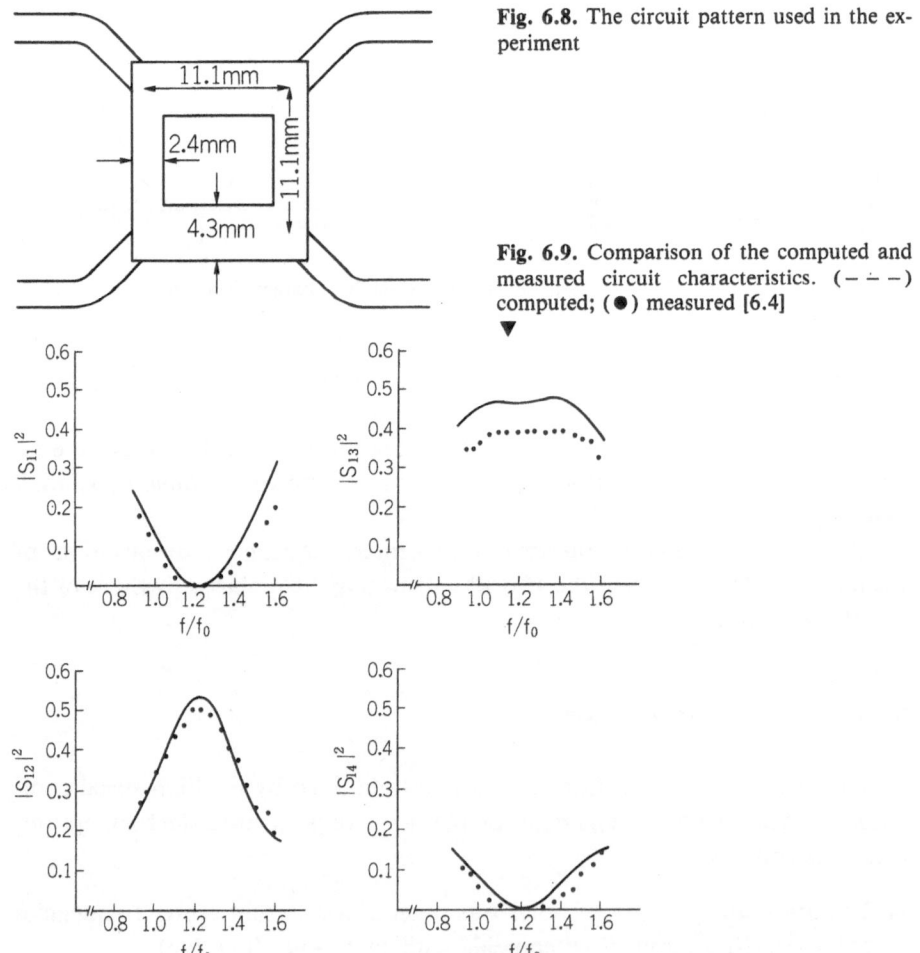

Fig. 6.8. The circuit pattern used in the experiment

Fig. 6.9. Comparison of the computed and measured circuit characteristics. $(-\cdot-)$ computed; ($\bullet$) measured [6.4]

A center conductor having the dimensions shown in Fig. 6.8 was sand-wiched by two ground conductors, with spacers of Rexolite 1422 ($d = 1.52$ mm, $\varepsilon_s = 2.53 \pm 0.03$). The center conductor was produced by a photoetching technique. In the analysis described in Chap. 5, the open-circuit boundary was assumed to exist at the periphery of the circuit. Actually, how-ever, the electromagnetic field extends outward due to the fringe effect. One method to take this effect into account in the analysis is to shift all the circuit boundaries outward by $0.442\,d$, in this case by 0.67 mm (Sect. 2.7 or [6.6]). The circuit dimensions shown in Fig. 6.8 were chosen so that the corrected pattern agrees with Fig. 6.7.

The computed and measured circuit characteristics are shown in Fig. 6.9 by solid curves and dots, respectively. They show good agreement with each other, except for a certain loss (approximately -0.7 dB) found in the measured value of $|S_{13}|^2$.

6.4 Computer Time

A high-speed analysis technique is a premise for the trial-and-error optimum design of a microwave integrated circuit. In the present case, the required computer time for analyzing one circuit pattern shown in Fig. 6.5 for a single frequency was about 2.5 s when 10×10 modes were considered in each seg-ment; a Hitachi computer HITAC 8800 was used. However, by decreasing the number of considered modes to a reasonable one (for example, 8×3), the re-quired time will easily be reduced to below 0.7 s. On the other hand, when we used the contour-integral method described in Chap. 3, the estimated time re-quired for the corresponding accuracy was about 5 s.

Generally speaking, comparison of the two analysis methods is rather dif-ficult. However, in many cases reduction of the computer time by almost one order of magnitude is possible by the segmentation method without great loss of accuracy as compared with the contour-integral method.

6.5 Summary

The principle and computation procedure of the trial-and-error, designer-as-sisted synthesis of a planar circuit using the segmentation method as the design tool has been described. To show the usefulness of the proposed meth-od, its application to the optimum design of a 3-dB hybrid circuit has been demonstrated.

The application of the same principle to the short-boundary planar circuit, which has mainly been dealt with by contour-integral analysis (Chap. 4), will be another interesting task. In the following chapter, a fully computer-orient-ed version of the synthesis will be presented.

7. Fully Computer-Oriented Synthesis of Optimum Planar Circuit Pattern

In contrast to the relatively primitive trial-and-error synthesis techniques described in the preceding chapter, a fully computer-oriented synthesis of the optimum planar circuit pattern is described now. A 3-dB hybrid circuit is again considered as an example, but the pattern is somewhat deformed from ladder-type to "ring-type" to facilitate mathematical expression of the circuit pattern. In the synthesis process, the contour-integral method and Powell's method are used for the circuit analysis and the optimization, respectively. The synthesized optimum patterns are given in normalized curves and parameters which can be used directly in practical circuit design. The validity of the theory is confirmed by experiment.

7.1 Background

As described in the preceding chapter, a branch-line 3-dB hybrid (Fig. 7.1a) consists of four stripline sections (arms) having length l equal to $\lambda_0/4$ and characteristic impedance equal to Z_0 and $Z_0/\sqrt{2}$, where λ_0 and Z_0 denote the center wavelength (reduced by dielectric material) and the impedance of the external striplines, respectively [7.1]. However, in some cases (for example, when the frequency becomes higher) the line widths W and W' become

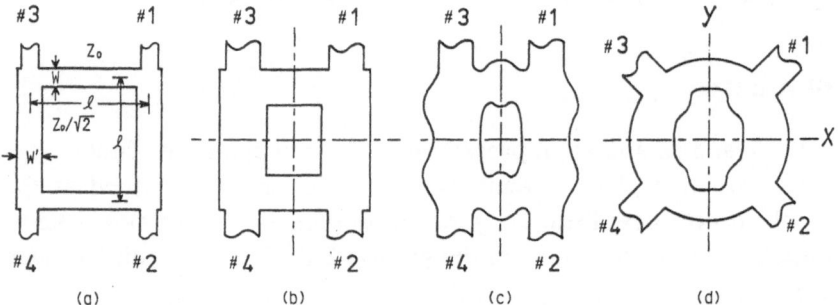

Fig. 7.1a–d. Ladder-type and ring-type 3-dB hybrids: (a) the distributed-constant model (one-dimensional stripline model) of a branch-line 3-dB hybrid, (b) a branch-line hybrid consisting of wide stripline sections, (c) an arbitrarily shaped planar 3-dB hybrid, (d) a planar 3-dB hybrid having a circular periphery (3-dB hybrid ring) [7.2]

comparable to the arm length (Sect. 6.2.2). In such a case the circuit pattern must be designed by planar circuit approach.

The optimum design of a circuit, as shown in Fig. 7.1b, based upon the planar circuit concept, has been described in the preceding chapter. It employed a computer-aided, but man-operated trial-and-error approach. The results obtained, however, were far from being general because the boundary of the circuit pattern was assumed to consist of straight lines, as shown in Fig. 7.1b.

When such a circuit is dealt with as a planar circuit, the circuit pattern should not necessarily consist of straight lines. We can take advantage of the larger degree of freedom inherent to the planar circuit pattern and use more arbitrary shapes such as shown in Fig. 7.1c to realize better wideband hybrid characteristics. However, in such a case the excessively large freedom in the circuit shape often makes it rather difficult to determine the optimum circuit pattern uniquely.

In this chapter, a fully computer-oriented synthesis of the optimum circuit pattern of such a hybrid circuit is described [7.2]. To prevent the possible difficulty resulting from the excessively large freedom as described above, the external periphery of the circuit pattern is assumed to be circular (Fig. 7.1d). The external diameter, position of external ports, and the shape of the internal periphery are adjusted to give the best wideband hybrid characteristics. The synthesized circuit patterns will be given as normalized curves for five values of relative linewidth: $W/\lambda_0 = 0.1$, 0.09, 0.08, 0.07, and 0.06.

The measured frequency characteristics of the optimized circuit patterns will be shown to confirm the validity of the theory (Sect. 7.5). To improve the hybrid characteristics further, some additional external circuits are proposed, and experimented with (Sect. 7.6). As a result, a bandwidth much wider than that of a conventional branch-line hybrid is achieved.

7.2 Method of Synthesis

7.2.1 Outline of Synthesis Process

The circuit pattern is defined first by several (in the present case, nine) real scalar variables. The frequency characteristics of a "starting" circuit pattern are computed, and the obtained characteristics are evaluated by means of an appropriately defined evaluation function which gives a real, scalar figure of merit. Next, the circuit pattern is modified a little according to an algorithm so that the figure of merit decreases. In other words, we search for the optimum point in the multidimensional space of the pattern variables. Such a "search" is continued until the figure of merit reaches its minimum.

In the actual synthesis of the optimum circuit pattern, the following choices of the computational methods and parameters are important:

1) choice of the method for analyzing the frequency characteristics of a given circuit,
2) choice of the pattern variables,
3) choice of the evaluation function,
4) choice of the algorithm for modifying the circuit pattern to minimize the figure of merit.

The above problems will be discussed in Sects. 7.2.2 – 5.

7.2.2 Analysis of Frequency Characteristics

The contour-integral method (Chapt. 3) is used because of its flexibility for an arbitrary circuit pattern. The drawbacks are the relatively long computer time and the fact that the derivatives of the figure of merit with respect to the pattern variables cannot be computed analytically. These drawbacks give constraints to the choices in Sects. 7.2.3 – 5.

7.2.3 Pattern Variables

The circuit pattern has a double symmetry with respect to x and y axes, as seen in Fig. 7.1d. Therefore, the entire circuit pattern is determined if the "quarter circuit" shown in Fig. 7.2 is given. The shape of the outer periphery of the quarter circuit is determined by two parameters: the radius r_0 and the position of external port θ_0. The shape of the inner periphery is more arbitrary. In the present analysis, we use n radial coordinates r_i ($i = 1, 2, \ldots, n$) for $\theta_i = 90° \times (i - 1)/(n - 1)$, as shown in Fig. 7.2. (In the actual synthesis, $n = 7$; see Sect. 7.3.1.) Thus the entire circuit pattern can be determined by $(n + 2)$ parameters: $r_0, \theta_0, r_1, \ldots, r_n$.

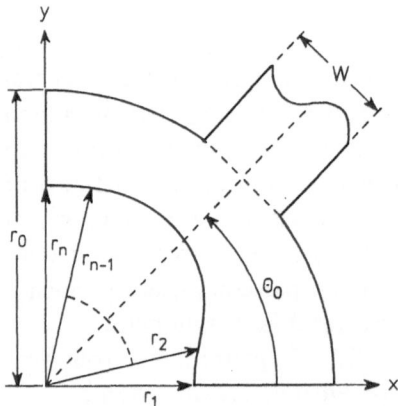

Fig. 7.2. A "quarter circuit" of the hybrid ring and the pattern variables used in the synthesis

In the actual analysis, the radial coordinates of the inner periphery for arbitrary values of θ are needed. To obtain these, interpolation between r_i's is made, by using a trigonometric series expansion

$$R(\theta) = C_0 + C_1 \cos 2\theta + C_2 \cos 4\theta + \ldots + C_{n-1} \cos 2(n-1)\theta, \tag{7.1}$$

where $C_0 \ldots C_{n-1}$ can be determined uniquely by $r_1 \ldots r_n$.

7.2.4 Evaluation Function

The choice of the evaluation function is the most critical step in the entire synthesis process. The properties to be taken into account in defining the evaluation function are as follow.

1) Do the following 3-dB hybrid characteristics hold for the scattering parameters S_{ij} in a wide frequency band?

$$|S_{11}|^2 \cong 0, \quad |S_{12}|^2 \cong 0.5,$$
$$|S_{13}|^2 \cong 0, \quad |S_{14}|^2 \cong 0.5. \tag{7.2}$$

2) Are the frequency characteristics symmetrical with respect to the center frequency f_0? That is, for a frequency deviation δf

$$|S_{1i}(f_0 - \delta f)|^2 = |S_{1i}(f_0 + \delta f)|^2, \quad i = 1, 2, 3, 4. \tag{7.3}$$

A general form of the evaluation function which approaches zero when (7.2, 3) are satisfied over a frequency band $(f_0 - \Delta f) \sim (f_0 + \Delta f)$ may be

$$F = F_1 + F_2, \quad \text{where} \tag{7.4}$$

$$F_1 = \sum_{i=1}^{4} \int_{f_0 - \Delta f}^{f_0 + \Delta f} \varrho_i(f) [|S_{1i}(f)|^2 - C_i]^m df, \tag{7.5}$$

$$F_2 = \sum_{i=1}^{4} \int_{0}^{\Delta f} \kappa_i(f') [|S_{1i}(f_0 + f')|^2 - |S_{1i}(f_0 - f')|^2]^{m'} df', \tag{7.6}$$

$$C_1 = C_3 = 0, \quad C_2 = C_4 = 0.5, \tag{7.7}$$

these F_1 and F_2 corresponding to the above two conditions (7.2, 3). In (7.5, 6), $\varrho_i(f)$ and $\kappa_i(f')$ are weighting functions, whereas exponents m and m' give nonlinearity to the evaluation.

However, in the actual computation, analysis of the circuit characteristics for a single frequency requires typically four seconds. Therefore, the integrals in (7.5, 6) must be replaced by a summation for a finite number of (practically, at most, several) frequencies. After some trials, it was found that three equally spaced frequencies $f_{-1}, f_0,$ and f_{+1} (Fig. 7.3) give satisfactory results,

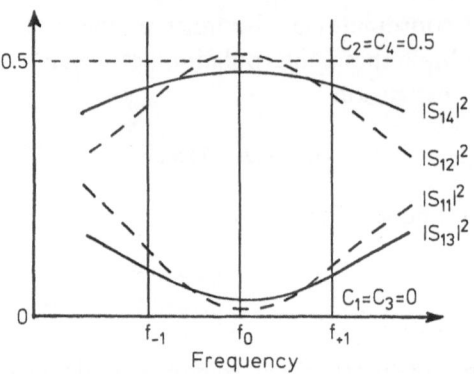

Fig. 7.3. Sampling frequencies used in the evaluation of the hybrid characteristics

and that a good convergence is obtained when $m = m' = 2$. Thus (7.5, 6) are finally much simplified as

$$F_1 = \sum_{i=1}^{4} \sum_{n=-1}^{+1} a_{in}[\,|S_{1i}(f_n)|^2 - C_i]^2, \tag{7.8}$$

$$F_2 = \sum_{i=1}^{4} b_i[\,|S_{1i}(f_{+1})|^2 - |S_{1i}(f_{-1})|^2]^2, \tag{7.9}$$

where a_{in} and b_i denote weighting parameters replacing $\varrho_i(f)$ and $\kappa_i(f')$.

7.2.5 Algorithm for Optimization of Circuit Pattern

Among various mathematical techniques for optimization, Powell's method [7.3] seems to be best suited to the present problem, in which the derivatives of the evaluation function with respect to pattern variables cannot be given analytically.

Powell's method has several versions; one used in the present synthesis will be described briefly in a comprehensive manner. For simplicity, we assume that only two independent variables are involved; that is, we are to find out the point Q which minimizes $f(x, y)$, as shown in Fig. 7.4a, under the condition that $(\partial f/\partial x)$ and $(\partial f/\partial y)$ cannot be computed. In the Powell's method, the "search" for the optimum point Q is performed for one variable (or a linear combination of variables) at a time, as shown in Fig. 7.4b:

1) starting from $P_0(x_0, y_0)$, x is optimized for $y = y_0$ (const) to obtain $P_1(x_1, y_0)$;
2) starting from P_1, y is optimized for $x = x_1$ (const) to obtain $P_2(x_1, y_2)$;
3) starting from P_2, (x, y) is optimized along a new vector $Z = X + Y$ to obtain P_3.

The above three steps [in an N-dimensional space, $(N+1)$ steps] constitute the first course. In the second course, usually y and Z directions are searched

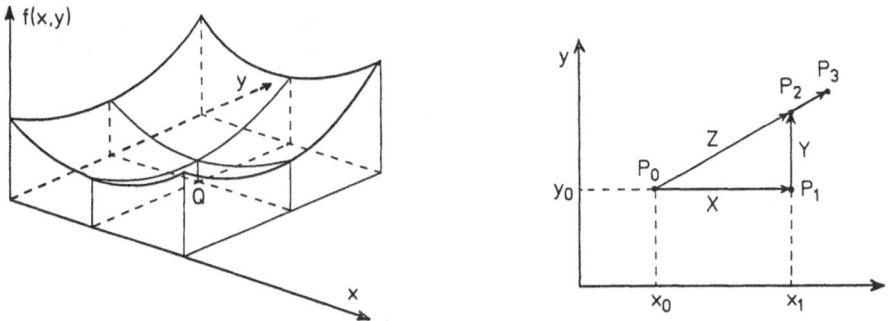

Fig. 7.4a, b. Powell's method: (**a**) a two-dimensional space and the optimum point Q, (**b**) process of the search employed in the present synthesis

in the first and second steps, respectively, and a new composite direction U is searched in the third step. In the third course, Z and U directions and a new composite direction are searched, and so forth. Usually satisfactory convergence is achieved within several courses.

As to the optimization in one step (e.g., finding P_1 along the line $y = y_0$), various well-established methods such as quadratic interpolation or golden-section search [7.3] can be used. In the present synthesis, the latter method is mainly used.

7.3 Parameters and Computational Techniques in an Actual Example of Synthesis

7.3.1 Number of Pattern Variables

Seven variables were used to express the inner periphery ($n = 7$). Hence the total number of variables is 9, and the above search is performed in a nine-dimensional space.

7.3.2 Reduction of Computer Time Taking Advantage of Double Symmetry

In the actual computations, the computer time can be reduced appreciably by taking advantage of the double symmetry of the circuit pattern. Instead of analyzing an entire four-port circuit as shown in Fig. 7.1d, we compute the complex reflection coefficients at the external port of the three-port quarter circuit, as shown in Fig. 7.5, for the following four terminating conditions:

1) with Ports A and B both open,
2) with Port A open and Port B grouped,

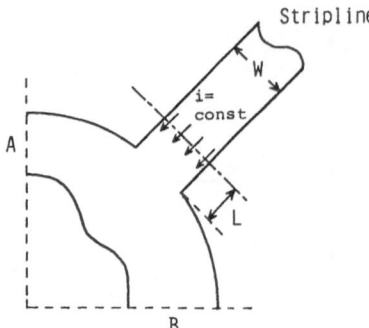

Stripline

Fig. 7.5. Definitions of Ports A, B and the external port where current uniformity can be assumed

3) with Port A grounded and Port B open,
4) with Ports A and B both grounded.

Scattering parameters S_{11}, S_{12}, S_{13}, and S_{14} of the entire four-port circuit are given as simple linear combinations of the complex reflection coefficients of the quarter three-port circuit for the above four terminating conditions [7.4].

7.3.3 Number of Sampling Points Along Periphery

In the contour-integral analyses of the circuit characteristics to be performed in each step of the synthesis, M sampling points must be given along the periphery of the circuit pattern for the numerical integration of the wave equation (Chap. 3). In the preliminary and final syntheses (Sect. 7.3.6), $M = 40$ and $M = 46$, respectively, for the quarter circuit.

7.3.4 Parameters in Evaluation Function

After preliminary trials, parameters in (7.8, 9) were chosen as

$$f_{-1} = 0.9f_0, \quad f_{+1} = 1.1f_0, \tag{7.10}$$

$$\left.\begin{array}{l} a_{i0} = 10 \\ a_{i,-1} = a_{i,+1} = 1 \\ b_{i,n} = 2 \end{array}\right\}, \quad i = 1, 2, 3, 4. \tag{7.11}$$

7.3.5 Relative Widths of External Striplines

The relative width of external striplines W/λ_0 is assumed to be 0.1, 0.09, 0.08, 0.07, and 0.06 where λ_0 denotes the center wavelength reduced by dielectric material.

7.3.6 Assumption of Uniform Current at Ports

At coupling ports, the current density is more or less nonuniform along the width of the stripline. To facilitate the analysis, however, we hope to assume that it is uniform. This assumption is permitted by defining an appropriate "port" which is a certain distance L apart from the periphery (Fig. 7.5). The longer the distance L, the better the computational accuracy, but the longer the computer time.

Some preliminary analyses showed that when $L = W/2$, the error of $|S_{1i}|^2$ ($i = 1, 2, 3, 4$) due to the finiteness of L is less than 1%. Therefore, in the actual synthesis, to save overall computer time, the synthesis process was divided into the following two parts:

1) preliminary synthesis, in which $L = 0$ and $M = 40$;
2) final synthesis, in which $L = 0.5\,W$ and $M = 46$.

7.4 Results of Synthesis

7.4.1 Process of Optimization

The decrease of the figure of merit F (7.4) during the synthesis process for the case $W/\lambda_0 = 0.09$ is shown as an example in the upper part of Fig. 7.6. The

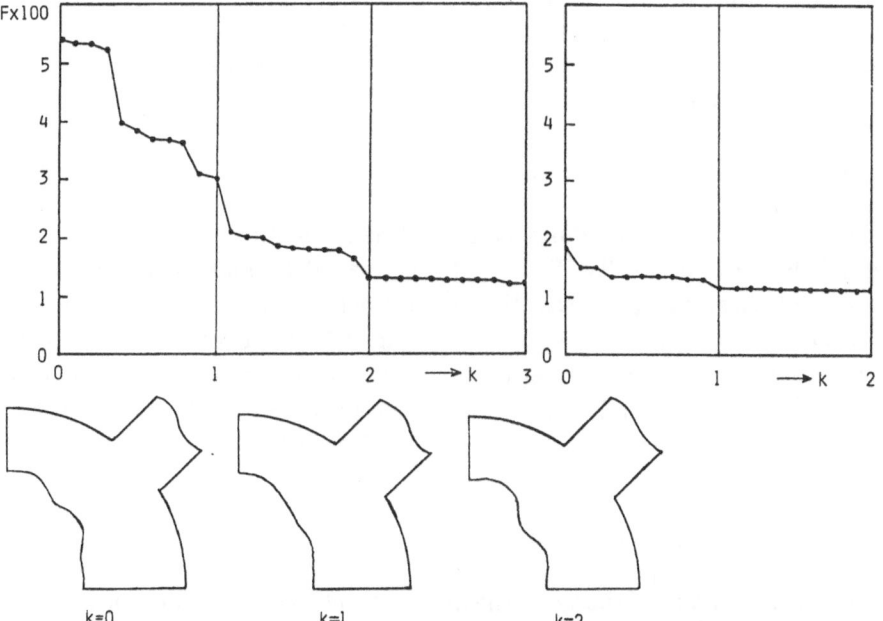

Fig. 7.6. The decrease of the figure of merit F and modification of the circuit pattern during the synthesis process [7.2]

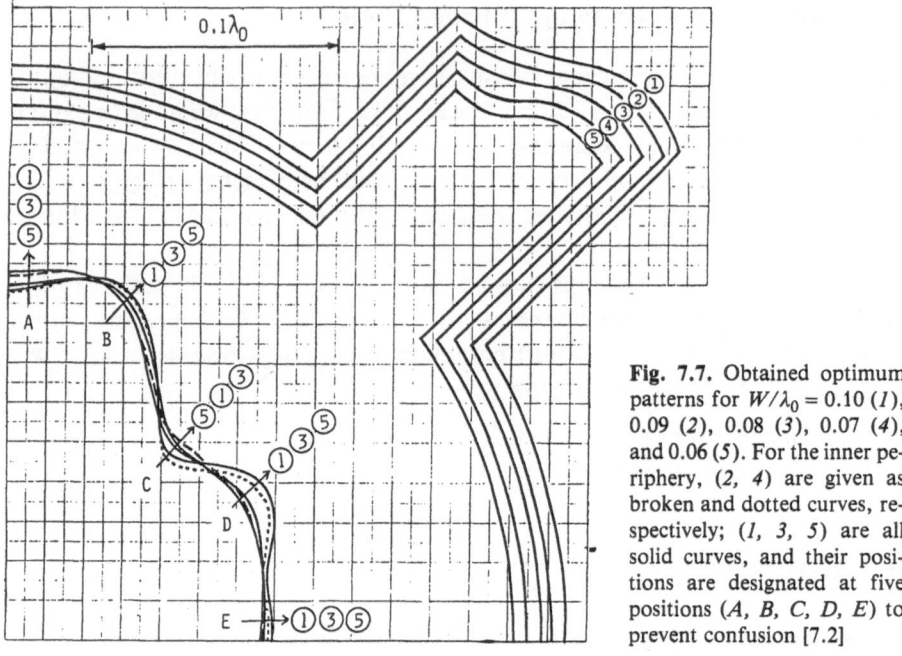

Fig. 7.7. Obtained optimum patterns for $W/\lambda_0 = 0.10$ (*1*), 0.09 (*2*), 0.08 (*3*), 0.07 (*4*), and 0.06 (*5*). For the inner periphery, (*2, 4*) are given as broken and dotted curves, respectively; (*1, 3, 5*) are all solid curves, and their positions are designated at five positions (*A, B, C, D, E*) to prevent confusion [7.2]

first three courses (shown in upper-left figure) constitute the preliminary synthesis in which $L = 0$, whereas the following two courses (upper-right figure) are the final synthesis in which $L = W/2$.

The corresponding circuit patterns are shown in the lower part of Fig. 7.6. The starting pattern was given empirically; actually some experience had been accumulated before this trial. The deformation after $\kappa = 2$ is not noticeable when drawn in this size.

7.4.2 Optimized Circuit Patterns

Figure 7.7 shows the optimum patterns finally obtained for $W/\lambda_0 = 0.10$, 0.09, 0.08, 0.07, and 0.06. These patterns are all normalized to the center wavelength λ_0 and can be used directly in practical circuit design. Note that the length 0.1 λ_0 is shown in the left top of Fig. 7.7. (For exact reproduction of these patterns, readers may refer to the optimized pattern variables tabulated in Ref. [7.2], Table 1.) The theoretical characteristics of these circuits are shown and compared with measured ones in Sect. 7.5.3.

7.5 Experimental Verification

To verify the validity of the synthesis theory, three hybrid circuits corresponding to cases $W/\lambda_0 = 0.10$, 0.08, and 0.06 have been fabricated, and their characteristics have been measured.

7.5.1 Circuit Design and Structure

A symmetrical, triplate structure is employed. The dielectric spacer is Rexolite 1442 ($\varepsilon_s = 2.53$) having thickness d of 1.45 mm (Fig. 7.8). The conductor thickness T ($\sim$0.1 mm) is neglected in the following design.

The circuit is designed first by assuming that a magnetic wall is present along the entire periphery of the circuit as well as the external striplines, in other words, neglecting the fringe effect as assumed in the synthesis. Under such an assumption, the equivalent linewidth W giving $Z_0 = 50\ \Omega$ can be determined, by using a relation

$$Z_0 = 377\ \Omega \times \frac{d}{2\sqrt{\varepsilon_s}\ W}, \tag{7.12}$$

to be 3.436 mm. The center frequencies f_0 corresponding to cases $W/\lambda_0 = 0.10$, 0.08, and 0.06 are then 5.50, 4.40 and 3.30 GHz, respectively.

The actual circuit patterns for the above three cases are shown in Fig. 7.9. To determine configurations of the ring-shaped portions of these circuits, the patterns given in Fig. 7.7 are modified (narrowed by an amount w along the entire periphery) to take into account the fringe effect. The value of w is given as [7.5]

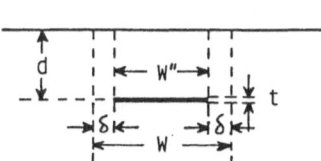

Fig. 7.8. Symbols used in the design of the experimental circuit and stripline

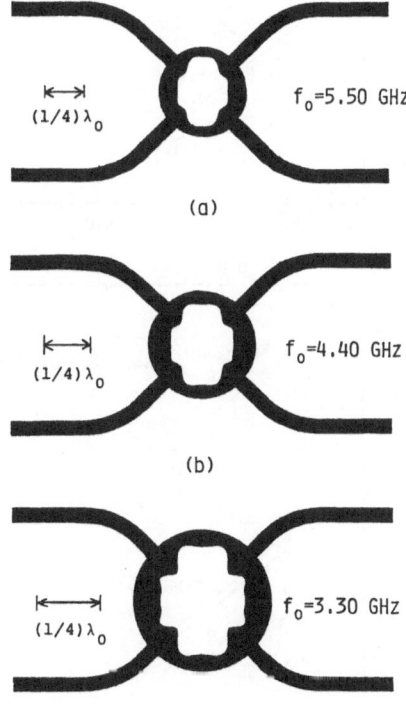

(a)

f_0=5.50 GHz

(b)

f_0=4.40 GHz

Fig. 7.9a – c. Three circuit patterns used in ▶ the experimental verification of the theory:
(a) $W/\lambda_0 = 0.10$ ($f_0 = 5.50$ GHz),
(b) $W/\lambda_0 = 0.08$ ($f_0 = 4.40$ GHz),
(c) $W/\lambda_0 = 0.06$ ($f_0 = 3.30$ GHz)

(c)

f_0=3.30 GHz

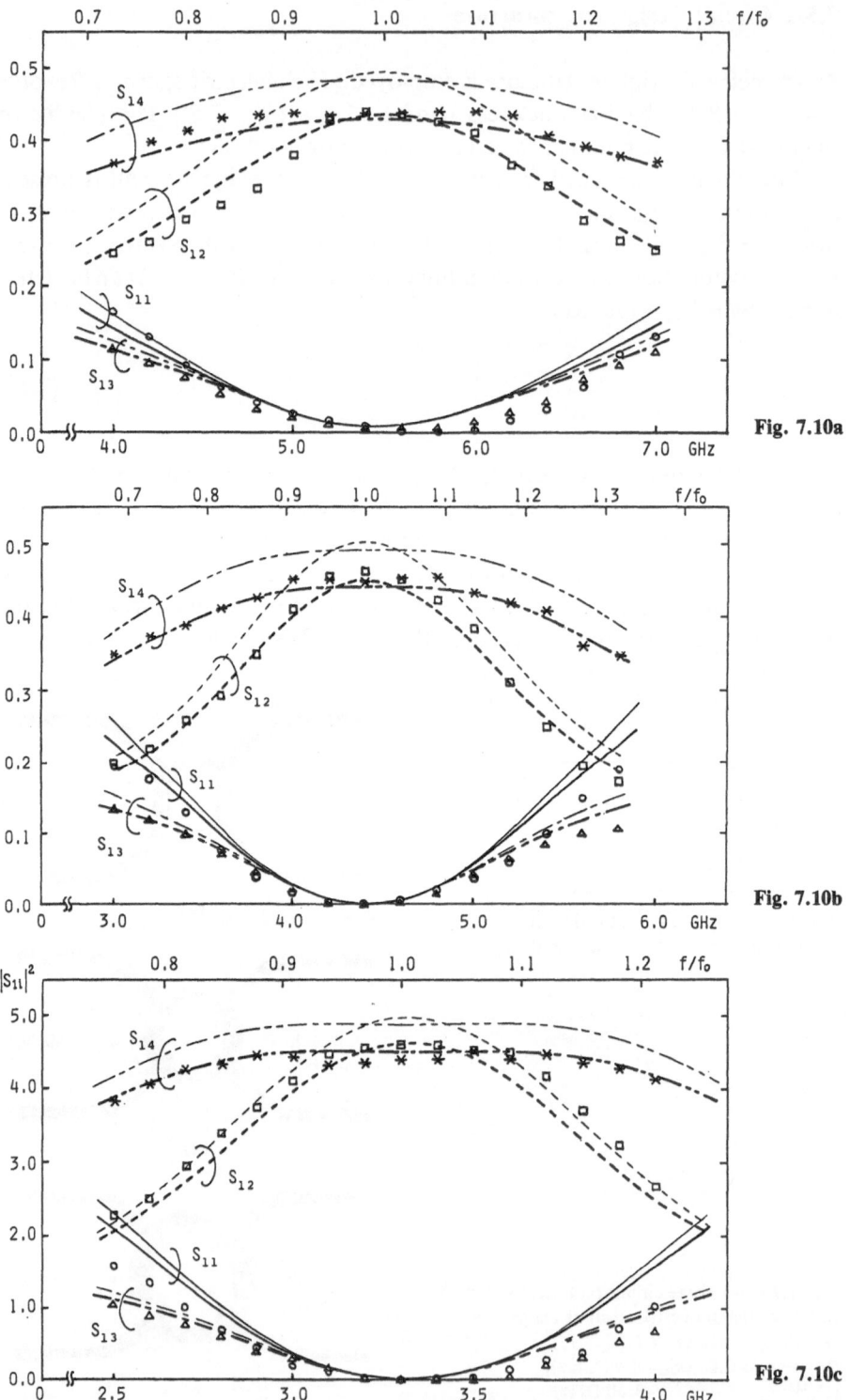

Fig. 7.10a

Fig. 7.10b

Fig. 7.10c

$$w = \frac{1}{\pi}\left[2d\ln 2 + \lambda S_1\left(\frac{4d}{\lambda}; 0, 0\right) - 2\lambda S_1\left(\frac{2d}{\lambda}; 0, 0\right)\right],\qquad(7.13)$$

where λ denotes the wavelength, and S_1 is a hypergeometric function defined as ([7.5] or Sect. 2.7)

$$S_1(x; 0, 0) = \sum_{n=1}^{\infty}\left(\sin^{-1}\frac{x}{n} - \frac{x}{n}\right).\qquad(7.14)$$

When $4d \ll \lambda$, (7.13) can be approximated as

$$w = 2d\ln 2/\pi = 0.442\,d,\qquad(7.15)$$

which corresponds to the static-field approximation. In the present case, (7.13) gives $w = 0.645$ mm, whereas (7.15) gives $w = 0.641$ mm. The difference between these is negligible.

The width of the stripline is also reduced from W to W'' (Fig. 7.8), which can be computed in two ways. One method is to use the formula lead by the Schwarz-Christoffel transform [7.6]; this leads to $W'' = 2.154$ mm. Another method is to assume that the amount of width reduction on one side δ (Fig. 7.8) is given approximately by (7.15). This leads to $W'' = 2.148$ mm. The difference between the two results is again below the fabrication error.

7.5.2 Result of Measurement

The measured frequency characteristics for the three experimental circuits are shown in Fig. 7.10a–c by small circles (S_{11}), squares (S_{12}), triangles (S_{13}), and asterisks (S_{14}). The thinner curves show the computed optimized characteristics, whereas the thicker ones are those obtained by correcting the thinner ones by circuit loss of 0.58 dB, 0.5 dB, 0.35 dB for the three cases, respectively. Those loss values have been obtained by measuring the insertion loss of a 55 mm long straight stripline section having OSM connectors (identical to those used in the hybrid) on both ends.

7.6 Further Improvement of Frequency Characteristics by Addition of External Circuits

The optimized frequency characteristics shown in Fig. 7.10, which themselves are appreciably better than conventional ones (Sect. 7.7), can further be

---

Fig. 7.10. Theoretical and measured frequency characteristics of the circuits shown in Fig. 7.9; Fig. 7.10a, b, c correspond to Fig. 7.9a, b, c. Measured data are shown by small circles (S_{11}), squares (S_{12}), triangles (S_{13}), and asterisks (S_{14}). For curves, refer to the text [7.2]

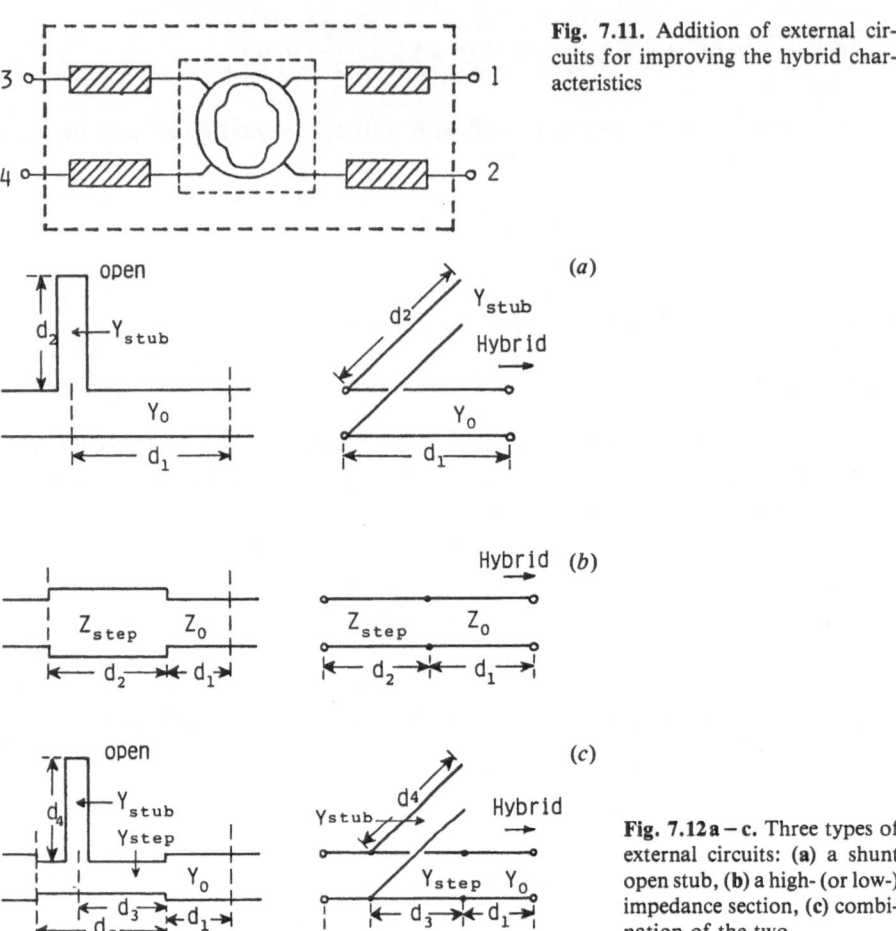

Fig. 7.11. Addition of external circuits for improving the hybrid characteristics

Fig. 7.12a – c. Three types of external circuits: **(a)** a shunt open stub, **(b)** a high- (or low-) impedance section, **(c)** combination of the two

improved by adding external circuits to all the four ports, as shown in Fig. 7.11. In the following, the optimum design of, and improvement achieved by, the addition of the external circuits will be described, though this topic is somewhat outside the research of planar circuits themselves. The original hybrid used in the design and experiment is the optimized one for $W = 0.1\,\lambda_0$, shown in Fig. 7.9a.

7.6.1 Types of External Circuits

The following three types of external circuits have been designed and experimented with:

1) a shunt open stub (Fig. 7.12a), which has once been used by *Riblet* [7.7] for the same band-widening purpose;

2) a high- (or low-) impedance section (Fig. 7.12b);
3) a combination of the above two schemes (Fig. 7.12c).

In Fig. 7.12, left and right figures show the actual circuit patterns and the equivalent circuits, respectively. The hybrid is connected to the terminals on the right side.

7.6.2 Optimization of Parameters

The Powell's method is again used to optimize the design parameters, that is,

1) Y_{stub}, d_1 and d_2 in Fig. 7.12a;
2) Z_{step}, d_1 and d_2 in Fig. 7.12b;
3) Y_{stub}, Y_{step}, d_1, d_2, d_3, and d_4 in Fig. 7.12c.

The computation of the frequency characteristics for the external circuits is based upon distributed-constant models. The original hybrid (Fig. 7.9a) is assumed to be unchanged.

Various evaluation functions were again tested in the preliminary design. Finally, however, one essentially identical to that given by (7.4, 7–9) was chosen and used in the design described in the following. The only difference was that the number of sampling frequencies was increased from 3 to 5; that is, characteristics were evaluated at $0.8 f_0$, $0.9 f_0$, f_0, $1.1 f_0$, and $1.2 f_0$.

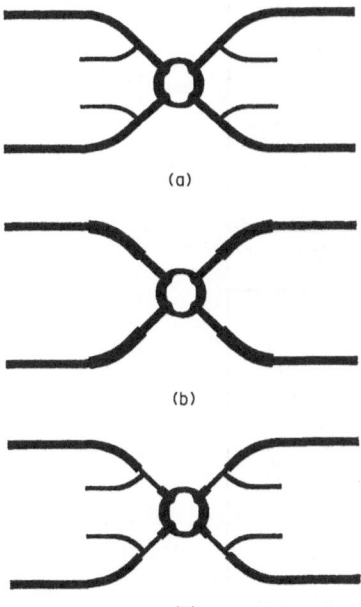

(a)

(b)

(c)

Fig. 7.13a–c. Three circuit patterns of the hybrid with optimized external circuits: (a) with shunt open stubs, (b) with low-impedance sections, (c) combination of shunt open stubs and high-impedance sections

7.6.3 Result of Optimization

Figure 7.13 shows three patterns of the hybrid with optimized external circuits. The obtained optimum design parameters are

1) $Y_{stub} = 0.6543\ Y_0$, $d_1 = 0.1770\ \lambda_0$, and $d_2 = 0.5027\ \lambda_0$ in Fig. 7.13a;
2) $Z_{step} = 0.7552\ Z_0$, $d_1 = 0.1702\ \lambda_0$, and $d_2 = 0.5076\ \lambda_0$ in Fig. 7.13b;
3) $Y_{stub} = 1.0431\ Y_0$, $Y_{step} = 0.6887\ Y_0$, $d_1 = 0.0\ \lambda_0$, $d_2 = 0.2268\ \lambda_0$,
   $d_3 = 0.1824\ \lambda_0$, and $d_4 = 0.4919\ \lambda_0$ in Fig. 7.13c.

The discontinuity effects (shift of reference planes) at branches and impedance steps have been taken into account in determining the actual circuit size.

7.6.4 Obtained Frequency Characteristics

Figure 7.14 shows the frequency characteristics obtained with the circuit pattern of Fig. 7.13b. The optimized theoretical characteristics are again shown

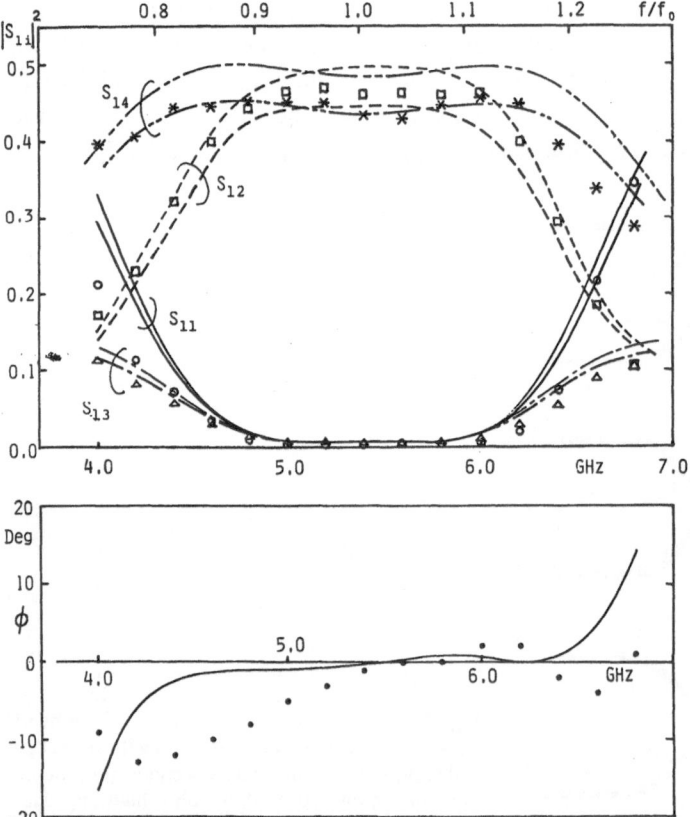

Fig. 7.14. Theoretical and measured frequency characteristics of the circuit shown in Fig. 7.13b. Symbols used are identical to Fig. 7.10 [7.2]

by relatively thin curves, and those corrected for the line loss (Sect. 7.5.2) by relatively thick ones. Small circles, squares, triangles, and asterisks again show S_{11}, S_{12}, S_{13}, and S_{14}, respectively. The lower part of Fig. 7.14 shows the differential phase angle

$$\phi = [\arg(S_{12}) - \arg(S_{14})] - 90° . \tag{7.16}$$

The difference between theory (solid curve) and experiment (dots) is below 10 degrees.

The theoretical and experimental characteristics obtained with the circuits of Fig. 7.13a, c are omitted because of space restrictions. The theoretical characteristics of these circuits show bandwidths a little wider than in Fig. 7.14. However, the experimental plots are scattered and show relatively large circuit loss (up to 0.05 in $|S_{12}|^2$ for Fig. 7.13a and 0.10 for Fig. 7.13c). It seems that the loss at the junction of stubs and in stubs themselves makes these circuits less practical at these frequencies.

7.7 Evaluation of the Synthesized Circuit Patterns

7.7.1 Comparison of Theory and Experiment

In Figs. 7.10, 14, the experimental and theoretical characteristics (corrected for the circuit loss) show good agreement within possible experimental error. The validity of the synthesis theory and the correctness of the computer programming seem to have been proved.

7.7.2 Comparison with Other Characteristics

Figure 7.15 shows the comparison of the following curves:

1) Broken curves: characteristics computed for the distributed-constant, branch-line hybrid, as shown in Fig. 7.1a;
2) Dash-dotted curves: characteristics of the hybrid circuit designed by the trial-and-error approach reported in Chap. 6;
3) Solid curves: theoretical characteristics obtained by the present synthesis for the case $W/\lambda_0 = 0.1$ (identical to thin curves in Fig. 7.10a);
4) Dotted curves: theoretical characteristics for $W/\lambda_0 = 0.1$ improved by the external low-impedance sections (identical to thin curves in Fig. 7.14).

Of the first three types of curves, the solid curves show the best performance, in particular for S_{12} and S_{11}. However, the improvement is not quite dramatic. (For example, if we define the bandwidth of S_{11} by $|S_{11}|^2 \leqslant 0.1$, its ratio is 1.00 : 1.02 : 1.20 for cases 1, 2, and 3.) This fact suggests that the synthesized

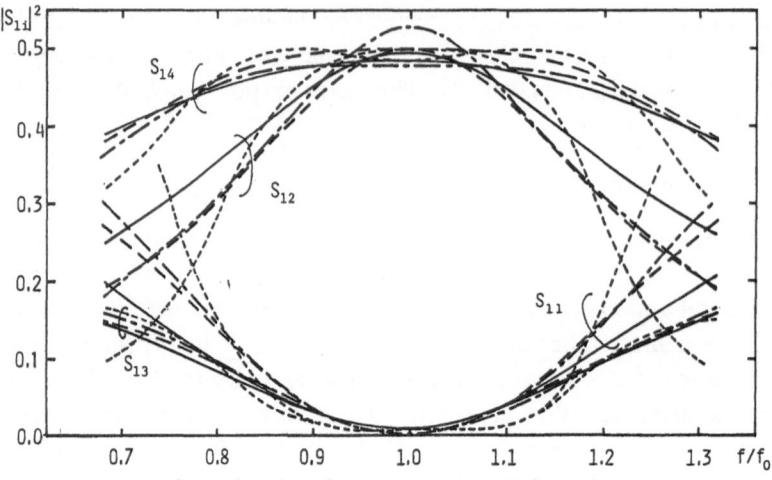

Fig. 7.15. Comparison of various frequency characteristics [7.2]

optimum patterns are still based upon the same principle as that of the distributed-constant, branch-line hybrid. In other words, the computer did not find out a novel hybrid principle in the course of the synthesis. If so, we should rather be pleased that the planar circuit approach has not only prevented the deterioration of the distributed-constant characteristics due to the widening of the circuit, but succeeded in improving those to some extent. The characteristics improved by the external circuits (dotted curves) have, of course, much better shapes than the first two types of curves (Fig. 7.15).

7.7.3 Comparison with Other Trials for Synthesizing Planar Circuits

In 1972 it was predicted that the significance of the planar circuit concept would be found in the synthesis (design) of circuit patterns in microwave integrated circuitry [7.8]. Since then, as described in Sect. 6.1, four papers dealing with the synthesis of planar circuits have appeared.

Grüner reported the conformal-mapping synthesis of a thin waveguide filter [7.9]. (A thin waveguide section can be regarded as a short-boundary planar circuit.) However, he described synthesis of only poles of transmission characteristics. Secondly, *Kato* et al. presented a fully computer-oriented iterative synthesis of open-boundary planar circuits having an impedance matrix with prescribed poles and residues [7.10]. However, this synthesis did not directly aim at any practical circuit design. Thirdly, *Okoshi* et al. described the trial-and-error design of a 3-dB hybrid having straight boundaries (Chap. 6 or [7.11]).

The novel points of the method described here [7.2] are (i) a practical wide-band circuit has been designed successfully, and (ii) Powell's method is used effectively, overcoming the difficulty of computing the derivatives of the evaluation function. The combination of the contour-integral analysis and Powell's optimization method described in this chapter offers a powerful tool in various design problems of planar circuit.

7.8 Summary

The fully computer-oriented synthesis of the optimum circuit pattern of a 3-dB hybrid ring based upon the planar circuit concept has been described. The synthesized optimum patterns are given in normalized curves which are convenient for practical use. The validity of the theory and the correctness of the computer programming have been proved by experiment. It has also been shown that the planar circuit approach not only can prevent the deterioration of the hybrid characteristics due to the widening of the circuit, but also bring forth characteristics somewhat better than the distributed-constant model. The obtained optimized characteristics can be further improved by the addition of simple external circuits to bring forth a much wider bandwidth.

8. Planar Circuits with Anisotropic Spacing Media

The analysis and design of a planar circuit having an anisotropic spacing medium are the subjects of this chapter. What is actually considered is a triplate-type planar circuit having ferrite spacers magnetized perpendicularly with respect to the circuit conductors. The analysis of an arbitrarily shaped ferrite planar circuit of this type is discussed first, aiming at the determination of the parameters of the equivalent multiport circuit.

In Sect. 8.5, the optimum design of a triplate-type stripline circulator is considered, aiming at wideband characteristics. This device must be dealt with as a planar circuit because it consists of a disk resonator, wide striplines, and tapered sections. It is shown that the planar circuit approach is a powerful tool in such a design. A 20 dB-isolation fractional bandwidth of 52% is achieved by this approach.

8.1 Background

In the preceding chapters, the space between conductors in a planar circuit has been assumed to be filled with isotropic material. Here, the analysis and design of a planar circuit having anisotropic spacing material are presented; the material actually considered is ferrite magnetized in the direction normal to the conductors. In particular, an arbitrarily shaped, triplate-type ferrite planar circuit is discussed. The discussion follows principally two papers by *Miyoshi* et al. [8.1, 2], which are the earliest contributions to the present problem. The analysis and optimum design of a stripline circulator [8.3] are the final target of the discussions. This device is now widely used in microwave integrated circuits, and must be dealt with, to be strict, as a planar circuit because it usually consists of a disk resonator, wide striplines, and tapered sections.

However, the discussion starts with the analysis of arbitrarily shaped, triplate ferrite planar circuits which include edge-guided mode devices [8.4, 5] as well as the disk-shaped circulator (Fig. 8.1). In particular, the circuit parameters of the equivalent multiport circuit are determined. To analyze ferrite planar circuits in general, two approaches are possible. One approach is based upon a contour-integral solution of the wave equation. In the other approach,

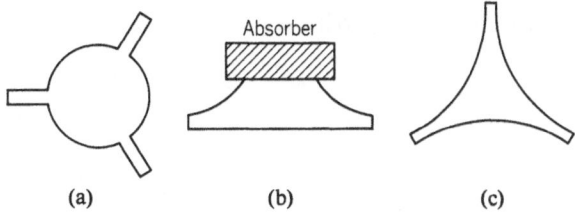

Fig. 8.1. Various ferrite planar circuits: (**a**) a disk-shaped circulator, (**b**) an edge-guided mode isolator, (**c**) an edge-guided mode circulator

the fields in the circuit are expanded in terms of orthonormalized eigenfunctions. Examples of such analyses are described and compared with experimental data to show the validity of the theory.

In Sect. 8.5, the optimum design of a triplate-type stripline circulator is considered. The optimum circuit shape for realizing a wideband operation is pursued, taking advantage of the wider freedom of the planar structure in circuit design.

8.2 Theories of Analysis

8.2.1 Basic Equations

We consider a triplate-type ferrite planar circuit, consisting of an arbitrarily shaped thin center conductor, two ferrite spacers, and two ground conductors. A dc magnetic field is applied normal to the ground conductors. The circuit is assumed to be excited symmetrically with respect to the upper and lower ground conductors. There are several coupling ports (Fig. 8.2) and the remainder of the periphery is assumed to be open circuited. The x, y coordinates and z axis, respectively, are set parallel and normal to the conductors. The thickness of the planar circuit is denoted by $2d$.

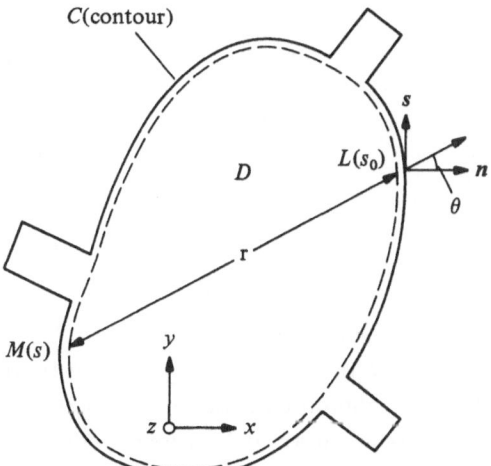

Fig. 8.2. Center conductor of a ferrite planar circuit and symbols used in the contour-integral equation

The tensor permeability of a magnetized ferrite is given as [8.6]

$$[\mu] = \begin{bmatrix} \mu & -j\kappa & 0 \\ j\kappa & \mu & 0 \\ 0 & 0 & \mu_0 \end{bmatrix}, \quad \text{where} \tag{8.1}$$

$$\mu = \mu_0 + \frac{\gamma^2 H_0 M_s}{\mu^2 H_0^2 - \omega^2}, \tag{8.2}$$

$$\kappa = \frac{\omega \gamma M_s}{\gamma^2 H_0^2 - \omega^2}. \tag{8.3}$$

Here μ_0 denotes the permeability of vacuum, γ the gyromagnetic ratio $[= -(1/2) g \mu_0 (e/m)$, where $g \doteq 2$ (depending upon material) and (e/m) is the charge-to-mass ratio of an electron], H_0 the dc bias magnetic field, M_s the saturation magnetization of the ferrite material, and ω the angular frequency of the signal.

When the spacing d is much smaller than the wavelength and ferrite spacers are homogeneous and linear, only the field components E_z, H_x, and H_y exist, with no variation in the z direction. In such a case, putting (8.1) into four equations constituting Maxwell's equations, i.e.,

$$\text{curl} E = -j\omega[\mu]H, \tag{8.4a}$$

$$\text{div}[\mu]H = 0, \tag{8.4b}$$

$$\text{curl} H = j\omega \varepsilon E, \tag{8.4c}$$

$$\text{div}\, \varepsilon E = \varrho, \tag{8.4d}$$

and assuming that $\varrho = 0$, we find, after some computations shown in Appendix A 8.1, that the following equation governs the electromagnetic field in the ferrite planar circuit:

$$(\nabla_T^2 + \omega^2 \varepsilon \mu_{\text{eff}}) V = 0. \tag{8.5}$$

Here

$$\nabla_T^2 = \frac{\partial^2}{\partial x^2} + \frac{\partial^2}{\partial y^2}, \tag{8.6}$$

$$\mu_{\text{eff}} = \frac{\mu^2 - \kappa^2}{\mu}, \tag{8.7}$$

and V given by $E_z \times d$ denotes the rf voltage of the center conductor with respect to the ground conductors. The effective permeability μ_{eff} is given by μ and κ, which are diagonal and off-diagonal components of the permeability tensor, as seen in (8.1). The sign of μ_{eff} depends upon the frequency and the internal magnetic field (8.2, 3, 7).

It is also shown in Appendix A8.1 that at coupling ports, the following boundary condition must apply:

$$j\frac{\kappa}{\mu}\frac{\partial V}{\partial s} - \frac{\partial V}{\partial n} = -j\omega\mu_{\text{eff}}di_n,\qquad(8.8)$$

where i_n is the surface current density normal to the periphery, and $\partial/\partial n$ and $\partial/\partial s$ denote the derivative normal to the periphery and the tangential derivative around the periphery, respectively.

In most parts of the circuit periphery where the coupling ports are absent, the current component normal to the periphery is zero, that is, $i_n = 0$. Actually, however, fringing magnetic fields are always present. A simple correction for this effect is to enlarge the periphery outwards by an amount of $0.477\,d\times K$ in advance of the analysis (Sect. 2.7). The coefficient K has been introduced by *Miyoshi* et al. [8.1]; it has been determined by comparing the measured resonant frequencies for various ferrite planar resonators with theoretical values computed by the Rayleigh-Ritz variational method, assuming that the circuits are lossless (Sect. 8.3.2).

8.2.2 Analysis Based on Eigenfunction Expansion

In the first approach to the analysis, we define first the Green's function $\mathscr{G}(x, y\,|\,x', y')$, as in Sect. 2.3.2, with which the rf voltage at a point in the circuit is expressed as

$$V = -j\omega d\mu_{\text{eff}}\oint_C \mathscr{G}(x, y\,|\,s_0)\,i_n(s_0)\,ds_0.\qquad(8.9)$$

It is assumed that $\mathscr{G}$ satisfies the following boundary condition along the contour C:

$$j\frac{\kappa}{\mu}\frac{\partial\mathscr{G}}{\partial s} - \frac{\partial\mathscr{G}}{\partial n} = 0.\qquad(8.10)$$

[Note that the dimension of $\mathscr{G}$ is different from that in (2.27).]

Next, we expand the Green's function in terms of the complex eigenfunctions ϕ_a which are derived from the following eigenvalue problem:

$$(\nabla_T^2 + \omega_a^2\varepsilon\mu_{\text{eff}})\phi_a = 0\quad(\text{in } D),\qquad(8.11)$$

$$j\frac{\kappa}{\mu}\frac{\partial\phi_a}{\partial s} - \frac{\partial\phi_a}{\partial n} = 0\quad(\text{on } C),\qquad(8.12)$$

$$\iint_D \varepsilon\phi_a\phi_b^*\,dS = \delta_{ab},\qquad(8.13)$$

where an asterisk denotes a complex conjugate. The rf voltage in the circuit can then be expressed, by using these eigenfunctions, as (Sect. 2.3.3)

$$V = j\omega d \oint_C \sum_{a=0}^{\infty} \frac{\phi_a \phi_a^*}{\omega_a^2 - \omega^2} (-i_n) ds_0 . \qquad (8.14)$$

The circuit parameters of the equivalent multiport can be computed by following the calculation shown in Sect. 2.3.2. We first define the rf voltage on a port and the total current flowing into a port, respectively, as

$$V_i = \frac{1}{W_i} \int_{W_i} V(s_i) ds_i , \qquad I_j = \int_{W_j} [-2i_n(s_j)] ds_j . \qquad (8.15)$$

Substituting (8.15) into (8.14), we have

$$V_i = \sum_{j=1}^{l} \left(\sum_{a=0}^{\infty} \frac{j\omega d}{2W_i W_j} \int_{W_i} \int_{W_j} \frac{\phi_a^*(s_j) \phi_a(s_i)}{\omega_a^2 - \omega^2} ds_i ds_j \right) I_j , \qquad (8.16)$$

where l is the number of the coupling ports. Thus the elements of the impedance matrix of the equivalent multiport are given as

$$Z_{ij} = \sum_{a=0}^{\infty} \frac{j\omega d}{2W_i W_j} \int_{W_i} \int_{W_j} \frac{\phi_a^*(s_i) \phi_a(s_j)}{\omega_a^2 - \omega^2} ds_i ds_j . \qquad (8.17)$$

The above equation tells us that the impedance matrix is not symmetrical, i.e., $Z_{ij} \neq Z_{ji}$, because the eigenfunction ϕ_a is generally a complex function. However, a relation $Z_{ij} = -Z_{ji}^*$ holds, which corresponds to the lossless condition of the circuit.

To derive the characteristics of a ferrite planar circuit by means of (8.17), we must solve the eigenvalue problem given by (8.11 – 13) repeatedly at different frequencies for a given bias magnetic field. This is because μ_{eff} (8.7) is a function of the operating frequency as well as the bias magnetic field. The method of this calculation will be described in Sect. 8.3.1.

8.2.3 Analysis Based on a Contour-Integral Equation

In the second approach to the analysis, we start from a contour-integral equation similar to (3.1), but modified a little to take the gyromagnetic property of the ferrite substrate into account.

We consider first the case when $\mu_{\text{eff}} > 0$. In such a case the starting integral equation is derived (Appendix A 8.2) as

$$V(s) = \frac{1}{2j} \oint_C \left[k \left(\cos\theta - j\frac{\kappa}{\mu} \sin\theta \right) H_1^{(2)}(kr) V(s_0) \right.$$
$$\left. - j\omega d\mu_{\text{eff}} H_0^{(2)}(kr) i_n(s_0) \right] ds_0 , \qquad (8.18)$$

where $H_0^{(2)}$ and $H_1^{(2)}$ are the zero-order and first-order Hankel functions of the second kind, respectively; i_n denotes the current density flowing outwards along the periphery; s and s_0 denote the distance along Contour C. The variable r denotes the distance between Points M and L represented by s and s_0, respectively, and θ denotes the angle made by the straight line from Point M to Point L and the normal at Point L (Fig. 8.2). If the current density i_n injected along the periphery is known, (8.18) becomes a Fredholm integral equation of the second kind in terms of the rf voltage.

The only difference between (3.1) and (8.18) is found in the first term of the right-hand side. The difference stems from the tensor representation of the permeability (Appendix A 8.2).

Next, we consider the case when $\mu_{\mathrm{eff}} < 0$. In this case, after calculations described in Appendix A 8.2, the following contour-integral equation is obtained:

$$V(s) = \frac{1}{2\mathrm{j}} \oint_C \left\{ k \left(\cos\theta - \mathrm{j}\frac{\kappa}{\mu}\sin\theta \right) [K_1(hr) - \mathrm{j}\,\pi I_1(hr)]\, V(s_0) \right.$$
$$\left. - \mathrm{j}\,\omega\, d\mu_{\mathrm{eff}} [K_0(hr) + \mathrm{j}\,\pi I_0(hr)]\, i_n(s_0) \right\} ds_0, \tag{8.19}$$

where K_1, I_1, K_0, and I_0 denote the first-order and zero-order modified Bessel functions of the second kind (K_1 and K_0) and first kind (I_1 and I_0), respectively, and

$$h = \omega\sqrt{\varepsilon\,|\mu_{\mathrm{eff}}|}. \tag{8.20}$$

The difference between (8.18) and (8.19) stems from the fact that, in the case $\mu_{\mathrm{eff}} > 0$, the Green's function is given as $\mathscr{G} = H_0^{(2)}(kr)/4\mathrm{j}$, whereas in the case $\mu_{\mathrm{eff}} < 0$ it is given as $\mathscr{G} = [K_0(hr) + \mathrm{j}\,\pi I_0(hr)]/2\pi$ (Appendix A 8.2).

8.3 Formulations for Numerical Computation and Examples of Calculation

The actual mathematical process of the numerical calculation of circuit characteristics and some examples of the calculation are described in the following four subsections for the eigenfunction-expansion method (Sects. 8.3.1, 2) and contour-integral method (Sects. 8.3.3, 4) [8.1].

8.3.1 Formulation for the Eigenfunction-Expansion Method

Because the impedance parameters Z_{ij} have already been given in terms of eigenvalues ω_a's and eigenfunctions ϕ_a's by (8.17), what remains is to solve

(8.11 – 13) to obtain ω_a's and ϕ_a's. For this purpose, as discussed in Sect. 3.7.2, the Rayleigh-Ritz variational method using a polynomial expansion is one of the appropriate methods.

The computational process will be described in a little more detail than in Sect. 3.7.5, which describes the formulation for the isotropic case. We note first that (8.11) is the Euler equation of the following functional:

$$I = \iint_D (|\nabla_T \phi|^2 - \omega^2 \varepsilon \mu_{\text{eff}} |\phi|^2) dS + j \frac{\kappa}{\mu} \oint_C \phi^* \frac{\partial \phi}{\partial s} ds . \tag{8.21}$$

Therefore, instead of solving (8.11), we try to find the complex function which minimizes the functional I.

We first express the function ϕ as a series

$$\phi = \sum_{i=1}^{M} c_i f_i , \tag{8.22}$$

where c_i's denote the complex expansion coefficients to be determined, f_i's are real basis functions, and M denotes the number of the basis functions to be considered. The stationary points of I can be obtained by solving M simulateous equations $\partial I / \partial c_i = 0$. This process leads to an eigenvalue problem

$$(A - \omega^2 \varepsilon \mu_{\text{eff}} B) c = 0 , \quad \text{where} \tag{8.23}$$

$$A_{ij} = \iint_D \nabla f_i \cdot \nabla f_j dS + j \frac{\kappa}{\mu} \oint_C f_i \frac{\partial f_j}{\partial s} ds , \tag{8.24}$$

$$B_{ij} = \iint_D f_i f_j dS , \tag{8.25}$$

and are $M \times M$ matrices, and c is a column vector defined as $c = [c_1, c_2, \ldots, c_M]^t$ (t: transpose).

Thus, the eigenvalue problem has been rewritten in a matrix form. Note that the values of μ, κ, and μ_{eff} are functions of both the bias magnetic field and frequency; this fact is an essential difference from the isotropic case. Since Matrix A is Hermitian and Matrix B is symmetrical and positive definite, the eigenvalues ω_a^2's are real. Eq. (8.23) can be solved easily by using a standard computer program if it is rewritten in a normalized form of the eigenvalue problem of a Hermitian matrix. To normalize the eigenfunctions, the coefficients obtained should be multiplied by [8.1]

$$\left(\varepsilon \sum_{i=1}^{M} \sum_{j=1}^{M} c_i^* c_j B_{ij} \right)^{-1/2} . \tag{8.26}$$

When the ferrite planar circuit has no coupling ports, i.e., in the case of an isolated resonator, the nontrivial condition leads to

$$\det (A - \omega^2 \varepsilon \mu_{\text{eff}} B) = 0 . \tag{8.27}$$

The resonant frequencies of the circuit are given by the roots of this equation as in the isotropic case.

8.3.2 Examples of Calculation

The characteristics of disk-shaped ferrite resonantors have well been understood [8.2, 7]. Therefore, in this section square and triangular resonators will be considered. In all of the following computations, a polynomial of order 5 is used to approximate the eigenfunctions; this gives a 21×21 matrix equation to be solved. Also, the ferrimagnetic material is assumed to be lossless with a saturation magnetization $4 \pi M_s = 1300$ G, specific permittivity $\varepsilon_s = 15.6$, and substrate thickness $d = 2$ mm.

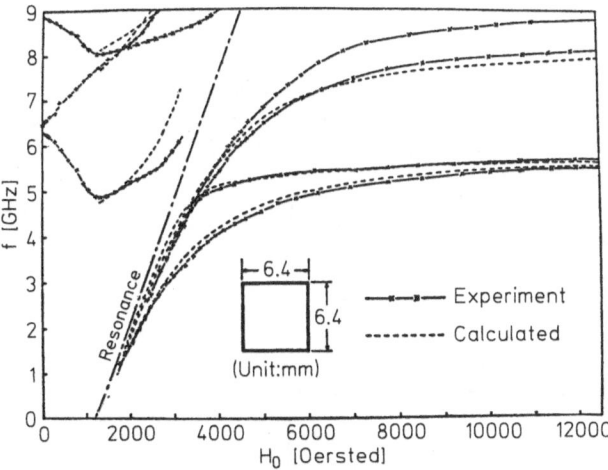

Fig. 8.3. Magnetic tuning characteristics of a square resonator. The broken curves have been obtained by taking fringing fields into account [8.1]

Figure 8.3 shows the magnetic tuning characteristics of a square resonator with dimensions of 6.4×6.4 mm^2. In computing the theoretical curves (broken curves), the effect of fringing fields is taken into account (Sect. 2.7). The measured resonant frequencies (solid curves) are found to be in good agreement with the theory, in particular at frequencies above the ferrimagnetic resonance. This is probably because the influence of the magnetic loss is smaller in this region. In the experiment, square ferrimagnetic substrates (25×25 mm^2) with $4 \pi M_s = 1300$ G, linewidth $\Delta H = 68$ Oe, $\varepsilon_s = 15.6$, and thickness $d = 2$ mm are used [8.1].

Figure 8.4 shows the calculated instantaneous distribution of the rf voltage in the square resonator for the fundamental mode. Equiamplitude (upper) and equiphase (lower) lines are shown for (a) $\mu_{\text{eff}} > 0$ at $H_0 = 1300$ Oe, and (b) $\mu_{\text{eff}} < 0$ at $H_0 = 2300$ Oe. The fields are found to rotate clockwise as in a disk

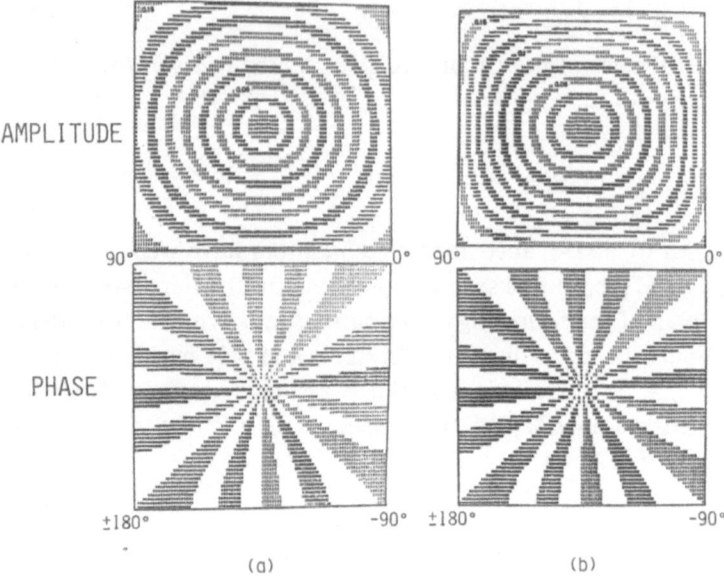

AMPLITUDE

PHASE

(a)   (b)

Fig. 8.4a, b. Computed instantaneous distribution of the rf voltage in a square resonator for the fundamental mode. Equiamplitude (*upper*) and equiphase (*lower*) lines are shown for (a) $\mu_{eff} > 0$ at $H_0 = 1300$ Oe, and (b) $\mu_{eff} < 0$ at $H_0 = 2300$ Oe [8.1]

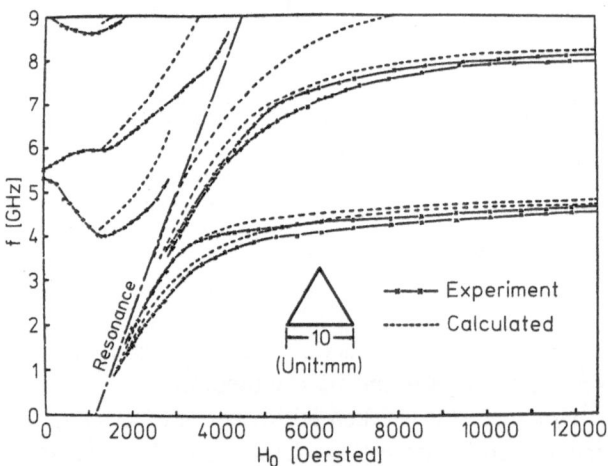

Fig. 8.5. Magnetic tuning characteristics of a triangular resonator [8.1]

resonator. It is also found that the field is concentrated along the periphery when $\mu_{eff} < 0$.

In the case of a triangular resonator with 10 mm sides, the magnetic tuning characteristics and the instantaneous rf voltage distributions for the fundamental mode are obtained as shown in Figs. 8.5, 6, respectively. These figures show almost the same resonant characteristics as obtained for a square resonator. In both cases, when $\mu_{eff} > 0$, the modes rotating clockwise and counterclockwise are both resonating modes. On the other hand, when

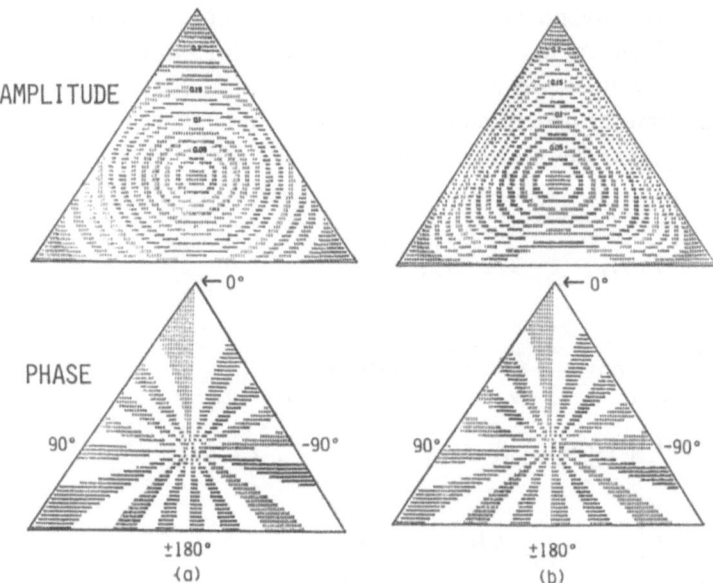

Fig. 8.6a, b. Computed instantaneous distribution of the rf voltage in a triangular circuit for the fundamental mode, for (a) $\mu_{\text{eff}} > 0$ at $H_0 = 1300$ Oe, and (b) $\mu_{\text{eff}} < 0$ at $H_0 = 2300$ Oe [8.1]

$\mu_{\text{eff}} < 0$, only the mode rotating clockwise is the resonating mode, and the field in the resonator is concentrated along the periphery.

8.3.3 Formulation for the Contour-Integral Method

In the numerical calculation of the contour-integral method, we divide the periphery into N incremental sections and set N sampling points defined at the

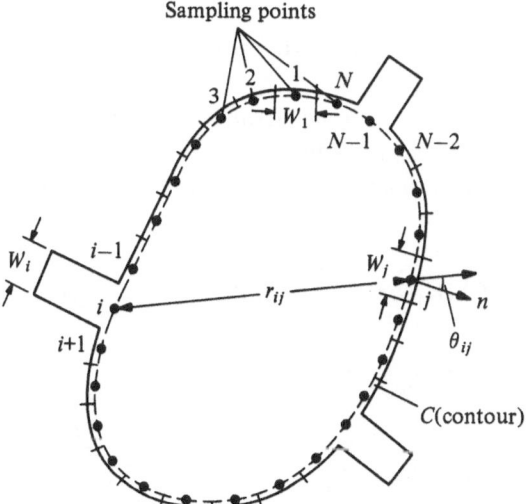

Fig. 8.7. Symbols used in the computer analysis

center of each section as shown in Fig. 8.7. The following computational process is essentially identical to the isotropic case (Sects. 3.2, 3); however, it will be reviewed to show the necessary modifications.

We consider first the case when $\mu_{\mathrm{eff}} > 0$. If we assume that the rf magnetic and electric field intensities are uniform across each section, (8.18) results in a matrix equation:

$$\sum_{j=1}^{N} u_{ij} V_j = \sum_{j=1}^{N} h_{ij} I_j, \qquad i = 1, 2, \ldots, N, \tag{8.28}$$

where

$$u_{ij} = \delta_{ij} - \frac{k}{2\mathrm{j}} \int_{W_j} \left\{ \cos\theta - \mathrm{j}\,\frac{\kappa}{\mu} \sin\theta \right\} H_1^{(2)}(kr)\,ds, \tag{8.29}$$

$$h_{ij} = \begin{cases} \dfrac{\omega\mu_{\mathrm{eff}}d}{4W_j} \displaystyle\int_{W_j} H_0^{(2)}(kr)\,ds, & i \neq j \\[2ex] \dfrac{\omega\mu_{\mathrm{eff}}d}{4} \left[1 - \dfrac{2}{\pi} \left(\ln\dfrac{kW_i}{4} - 1 + \gamma \right) \right], & i = j. \end{cases} \tag{8.30}$$

Here $\gamma = 0.5772 \ldots$ is Euler's constant, and $I_j = -2i_n W_j$ represents the total current flowing into the jth port. Equations (8.29, 30), giving u_{ij} and h_{ij}, have been derived in Sect. 3.3, assuming that the jth section is straight. From the above relations, the impedance matrix of the equivalent N port is given as

$$Z = U^{-1}H, \tag{8.31}$$

where U^{-1} denotes the inverse matrix to U.

When the circuit has no coupling port, i.e., $I_j = 0$, from the nontrivial condition of (8.28), we have

$$\det U = 0. \tag{8.32}$$

This equation also the resonant frequency gives in the case of a ferrite planar circuit.

Note that in the case $\mu_{\mathrm{eff}} > 0$, the only difference from the isotropic case is found in (8.29), i.e., the term $(-\mathrm{j}\kappa \sin\theta/\mu)$ is added.

When $\mu_{\mathrm{eff}} < 0$, the elements of matrices U and H in (8.31) are obtained, by comparing (8.18) and (8.19), as

$$u_{ij} = \delta_{ij} - \frac{h}{\pi} \int_{W_j} \left(\cos\theta - \mathrm{j}\,\frac{\kappa}{\mu} \sin\theta \right) [K_1(hr) - \mathrm{j}\pi I_1(hr)]\,ds, \tag{8.33a}$$

$$h_{ij} = \begin{cases} \dfrac{\mathrm{j}\omega\mu_{\mathrm{eff}}d}{2\pi} \dfrac{1}{W_j} \displaystyle\int_{W_j} [K_0(hr) + \mathrm{j}\pi I_0(hr)]\,ds, & i \neq j \\[2ex] -\dfrac{\mathrm{j}\omega\mu_{\mathrm{eff}}d}{2\pi} \left[\left(\ln\dfrac{hW_i}{4} + \gamma - 1 \right) - \mathrm{j}\pi \right], & i = j. \end{cases} \tag{8.33b}$$

8.3.4 Examples of Calculation

We assume again that the saturation magnetization $4\pi M_s = 1300$ G, the specific permittivity $\varepsilon_s = 15.6$, and the substrate thickness $d = 2$ mm.

The resonant frequencies of a disk-shaped circuit are computed first to check the computation accuracy. Since $|\det U| = 0$ in (8.32) can never be realized for real frequency due to the computation error, we define the frequency which gives the minimum of $|\det U|$ as the eigenvalue. The variation of $|\det U|$ is shown as a function of frequency in Fig. 8.8 for $N = 33$ at $H_0 = 4000$ Oe, which shows the first (4.35 GHz), the second (5.31 GHz), the third (6.05 GHz), the fourth (6.85 GHz), and the fifth (7.26 GHz) minima. By comparing these calculated eigenvalues with theoretical ones, which are the roots of

$$J'_n(ka) - \frac{\kappa}{\mu}\frac{nJ_n(ka)}{ka} = 0, \quad n = 0, \pm 1, \pm 2, \ldots, \tag{8.34}$$

$$I'_n(ha) - \frac{\kappa}{\mu}\frac{nI_n(ha)}{ha} = 0, \quad n = 1, 2, \ldots, \tag{8.35}$$

we find that the computation error is within 2.0% for the sampling point number of 33.

As the second example of the calculation, the computed characteristics of a Y-junction, disk-shaped circulator are shown in Fig. 8.9. It is assumed that the internal magnetic field is 3700 Oe, $N = 33$, and 50-Ω striplines are coupled to the circulator. In Fig. 8.9, the best circulator performance is obtained at frequencies a little below the ferrimagnetic resonance frequency, which is

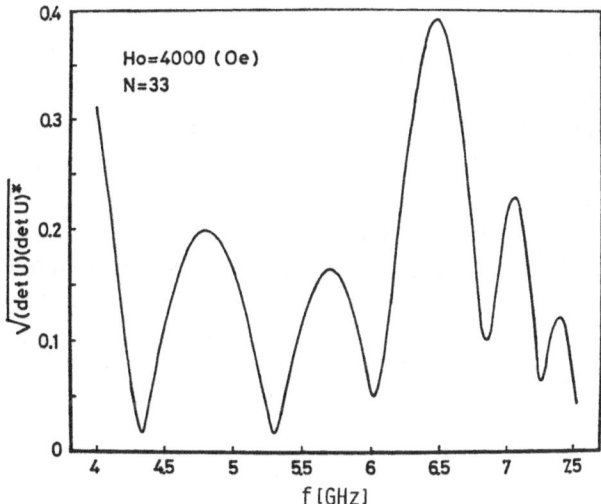

Fig. 8.8. The variation of $|\det U|$ as a function of frequency for a disk-shaped circuit at $H_0 = 4000$ Oe [8.1]

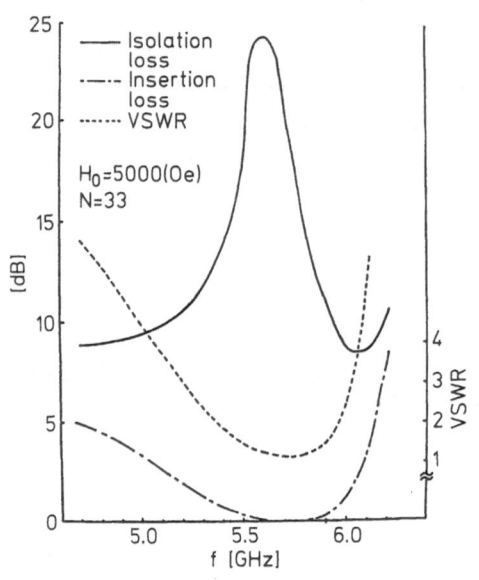

Fig. 8.9. Computed performance of a Y-junction, disk-shaped circulator [8.1]

Fig. 8.10. Computed rf voltage distribution, amplitude (*solid curve*) and phase (*broken curve*), along the periphery of a Y-junction, disk-shaped stripline circulator at the center frequency [8.1]

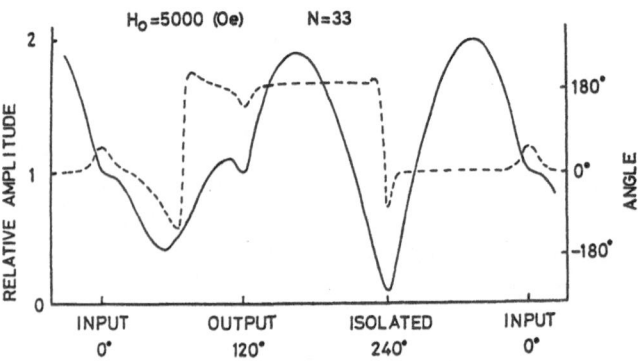

about 5.7 GHz in this case. On the other hand, the resonant frequencies of $+1$ and -1 modes, respectively, are 5.5 and 4.9 GHz.

Figure 8.10 shows the rf voltage distribution along the periphery of a Y-junction, disk-shaped circulator at the center operating frequency. The solid and broken curves show the relative amplitude and phase of the rf voltage, respectively. The distribution of the amplitude is not sinusoidal, but exhibits a deep minimum at the isolated port, a shallow minimum between the input and output ports, and a distorted field distribution in the vicinity of input/output ports. This is due to the influence of higher-order modes which results from the strong coupling to the external ports.

8.4 Comparison of the Eigenfunction-Expansion and Contour-Integral Methods

In this section, the application of the two methods to a triangular ferrite planar circuit is described to compare the results.

The shape and dimensions of the center conductor of the triangular ferrite planar circuit used in the calculation are shown in Fig. 8.11. The shaded portion (hexagon) is regarded as a planar circuit. The characteristics of the circuit are calculated for various applied magnetic fields and characteristic impedances Z_0 of the stripline. When the applied magnetic field is 5300 Oe, a circulator performance as shown in Fig. 8.12 is obtained above the ferrimagnetic resonance when $Z_0 = 30 \, \Omega$. The thinner curves have been obtained by the eigenfunction-expansion method using 21×21 matrices, the thicker curves by the contour-integral method with $N = 33$. The performance obtained in Fig. 8.12 can be explained by considering two fundamental rotating modes, i.e., modes rotating clockwise and counterclockwise. The principle of operation is thus the same as for a disk-shaped stripline circulator.

When the applied magnetic field is 3300 Oe, μ_{eff} is negative in the frequency range between 7.19 and 9.24 GHz. In this range the circuit functions as the so-called edge-guided mode circulator. Figure 8.13 shows the calculated

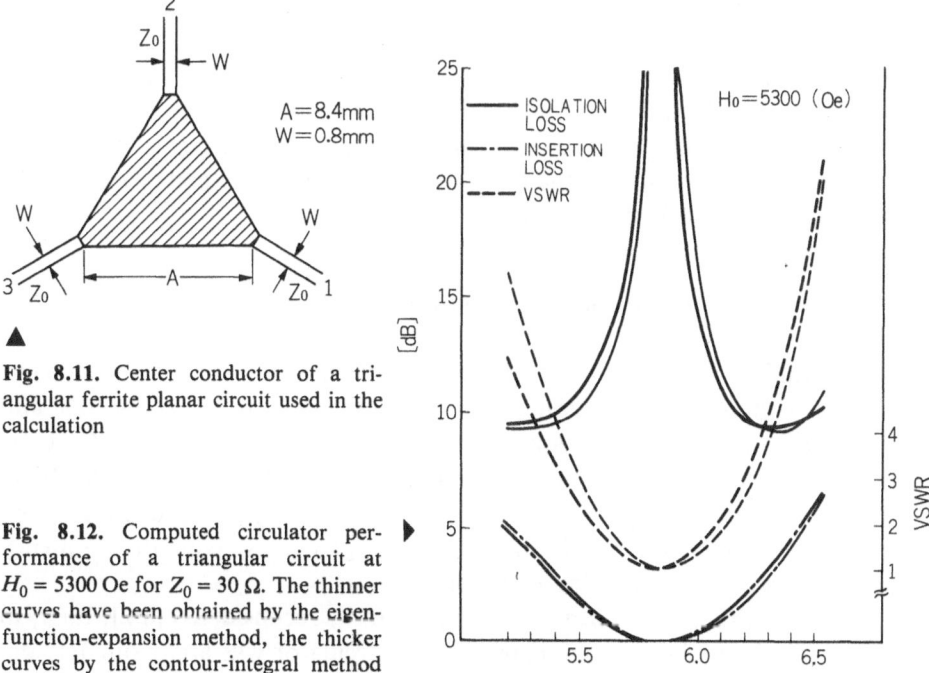

Fig. 8.11. Center conductor of a triangular ferrite planar circuit used in the calculation

Fig. 8.12. Computed circulator performance of a triangular circuit at $H_0 = 5300$ Oe for $Z_0 = 30 \, \Omega$. The thinner curves have been obtained by the eigenfunction-expansion method, the thicker curves by the contour-integral method [8.1]

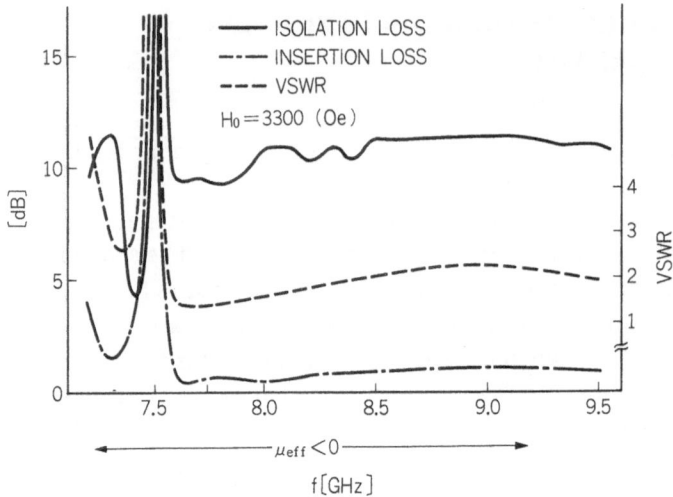

Fig. 8.13. Computed circulator performance of a triangular circuit at $H_0 = 3300$ Oe [8.1]

performance for $Z_0 = 50\,\Omega$. This type of circuit has not been fully exploited experimentally.

8.5 Optimum Design of Ferrite Planar Circuits

Upon the basis of the analysis techniques so far described, we next consider the synthesis (optimum design) of ferrite planar circuits [8.2]. The example to be considered is again a three-port circulator. After a short review of technical and historical backgrounds, we consider first, as a preparatory step, the optimum design of disk-shaped and triangular circulators, and then proceed to the design of a more arbitrary planar circulator.

8.5.1 Technical and Historical Backgrounds

The possibility of an octave-bandwidth design of a disk-shaped stripline circulator without external tuning elements was presented first by *Wu* and *Rosenbaum* in 1974 [8.8], whose analysis was based upon *Bosma*'s Green's-function method [8.3]. Their work was epochal in the design of wideband stripline circulators. Afterwards, *Ayter* and *Ayasli* [8.9] examined further Wu and Rosenbaum's design, and showed the design curves giving the frequency dependence of the circulation equation.

However, there remained two problems to be reconsidered [8.2]. First, in their analysis the impedance matrix of the circulator was approximately defined (after *Bosma*) by using the average voltage and current on each port,

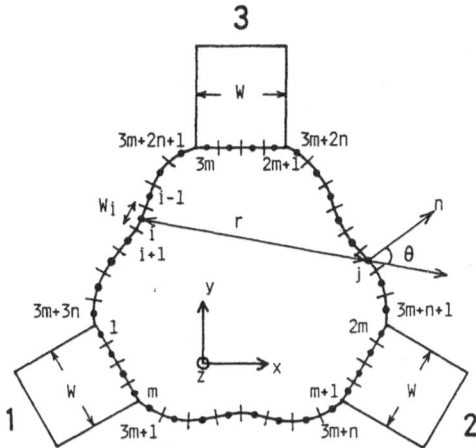

Fig. 8.14. Sampling points and symbols used in the numerical calculation of a planar circulator [8.2]

even when the junction coupling angle was large. Second, although the possibility of the octave-bandwidth operation was predicted, the upper limit of the 20-dB (for example) isolation bandwidth of the disk-shaped circulator was not shown quantitatively.

In the following discussions, therefore, following *Miyoshi* and *Miyauchi* [8.2], we first consider the upper limit of the 20-dB isolation fractional bandwidth of a disk-shaped circulator. Next, to find the optimum shape for wideband operation, a triangular circulator and its versions having various curved sides are considered.

8.5.2 Method of Numerical Analysis

In the course of the optimum circuit design, the process of circuit analysis is inevitable. In the following design, the contour-integral method described in Sects. 8.2.3 and 8.3.3, 4 is used throughout. Figure 8.14 shows an example of sampling-point arrangement along the periphery of a circulator and symbols used in the calculation.

8.5.3 Optimum Design of a Disk-Shaped Circulator

The contour of a disk-shaped circulator is not a perfect circle but has three straight sections at the ports as seen in the inset of Fig. 8.15. We call the angle subtended by these straight sections the "coupling angle".

The characteristics of disk-shaped circulator have been investigated in terms of the coupling angle. It has been found [8.8] that a coupling angle of 48 degrees gives the best fractional bandwidth. The computed performance of such a circulator is shown in Fig. 8.15. The 20-dB isolation fractional bandwidth is found to be about 43%, whereas the VSWR and the insertion loss are

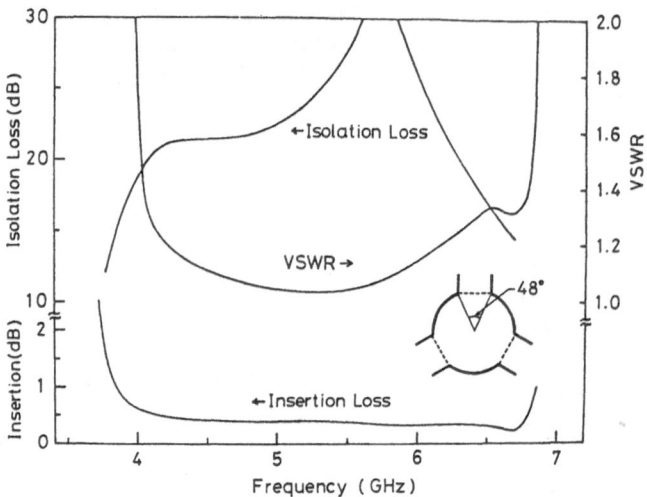

Fig. 8.15. Computed performance of a disk-shaped circulator having a coupling angle of 48 degrees [8.2]

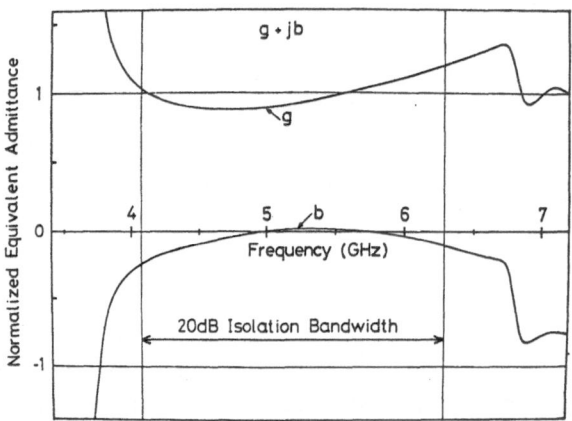

Fig. 8.16. Normalized equivalent admittance $g+jb$ of the disk-shaped circulator shown in Fig. 8.15 [8.2]

less than 1.3 and 0.6 dB, respectively, over that 43% frequency range. Other ranges of parameters are $0.58 < |\kappa/\mu| < 0.89$ and $2.09 > kR > 0.76$, where R is the radius of the disk.

Figure 8.16 shows the equivalent admittance [8.10] of the circulator computed from the admittance matrix components. The equivalent admittance is the input admittance of a circulator when all the ports are connected to matched external circuits. Figure 8.16 shows that the normalized equivalent admittance $g+jb$ is close to $1+j0$ over the 43% frequency range.

8.5.4 Optimum Design of a Triangular Circulator

Figure 8.17 shows the structure and parameters of a triangular circulator. In the numerical optimization of the circuit pattern, the four parameters ε_s, d_s,

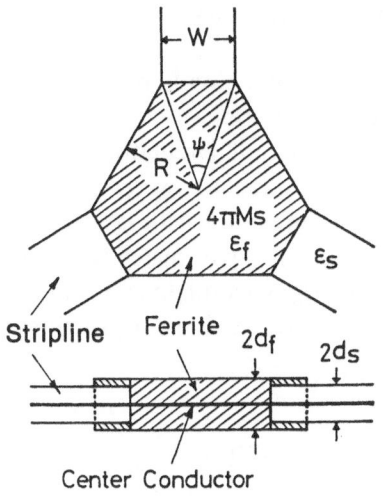

Fig. 8.17. Structure of a triangular circulator to be optimized [8.2]

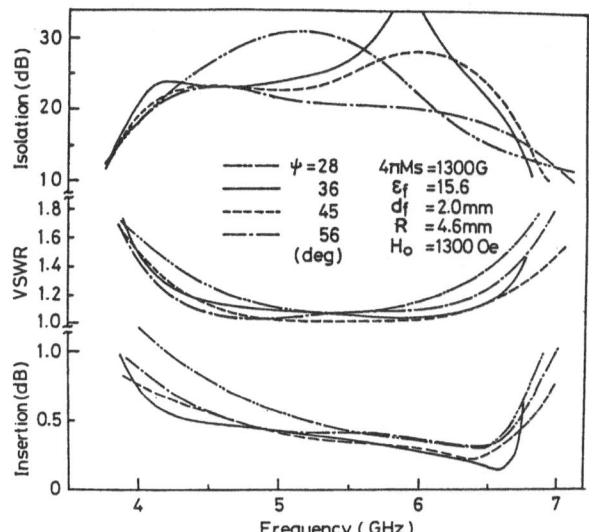

Fig. 8.18. Computed performances of the triangular circulators optimized in terms of ε_s and d_s for various coupling angles [8.2]

R, and ψ are optimized for a ferrite material with $4\pi M_s = 1300$ G, $\varepsilon_f = 15.6$, and $d_f = 2.0$ mm. In the circuit analysis 39 sampling points in total along the periphery and 5 sampling points on each port have been used.

Figure 8.18 shows the computed circulator performance obtained by optimizing ε_s and d_s for various coupling angles, where R is assumed to be 4.6 mm. The best 20-dB isolation fractional bandwidth is found to be about 50% for coupling angles between 35° and 45°. This value of fractional bandwidth is better than that of the disk-shaped circulator. *Helszajn* et al. also gave the same conclusion for circulators using disk-shaped and triangular planar resonators [8.11].

8.5.5 Modified Triangular Circulators Having Curved Sides

Next, the three straight sides of the optimized triangular circulator are modified to various curved sides to obtain better characteristics.

Figure 8.19 shows the isolation loss performance of circulators having various convex sides, where all the geometrical parameters except the side shape are kept the same as those of the triangular circulator with $\psi = 36°$. A portion of a circle is used to express the side shape. The parameter H in Fig. 8.19 denotes the maximum distance between the original straight side and the curved side. It is found in Fig. 8.19 that the larger H, the less the fractional bandwidth.

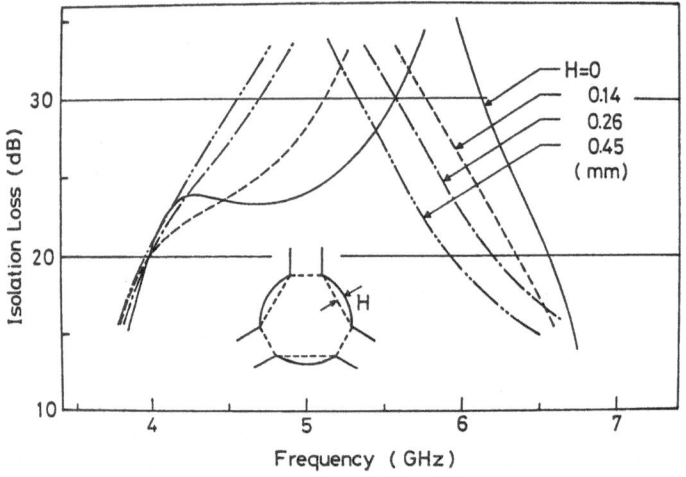

Fig. 8.19. Computed isolation loss performances of circulators with various convex sides [8.2]

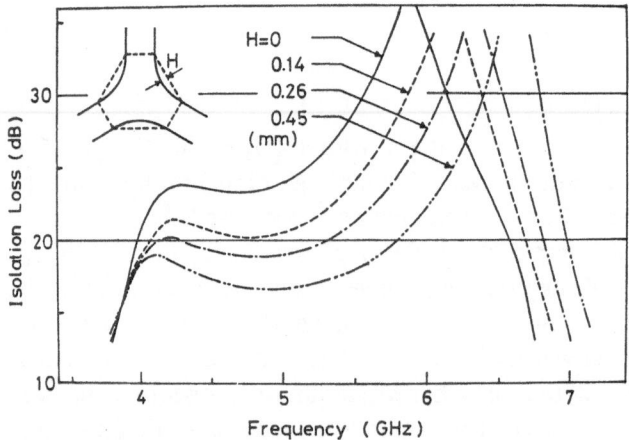

Fig. 8.20. Computed isolation loss performances of circulators with various concave sides [8.2]

The isolation loss performances of "concave" circulators are shown in Fig. 8.20. From this figure, a circulator having slightly concave sides is expected to give a little better fractional bandwidth if all the other parameters are optimized.

A trial-and-error approach gives us the best amount of the deformation. If we denote the radius of the inscribed circle of the original triangle by R, the best H/R value is found to be 0.057. The computed performance of such a circulator is shown by solid curves in Fig. 8.21 for $Z_1 = 11.26 \ \Omega$. The broken curves are for the best triangular circulator. The 20-dB isolation fractional bandwidth is about 52% for the former. Comparing the two curves, we see that the performance of the concave circulator is improved, especially at higher frequencies.

The normalized equivalent admittance of this circulator is shown in Fig. 8.22. The broken curves are again for the triangular circulator with $\psi = 36°$. It is found that the real and imaginary parts of the equivalent admittance

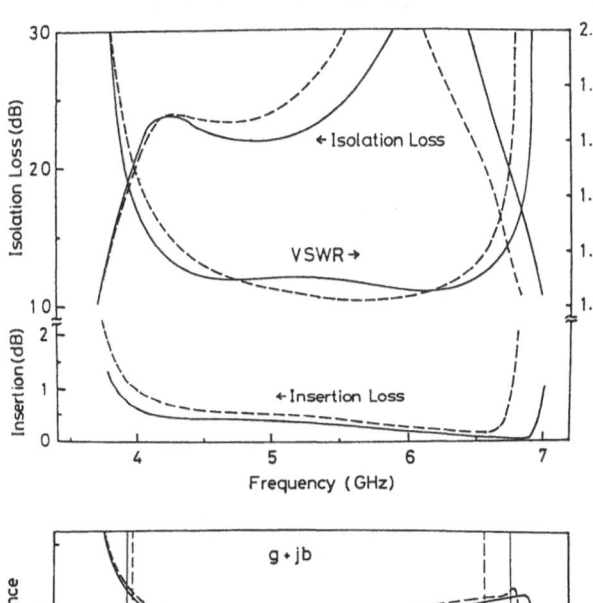

Fig. 8.21. Computed performance of the circulator having the optimized shape, i.e., a slightly concave sides with $H/R = 0.057$. The broken curves are for the triangular circulator with $\psi = 36$ [8.2]

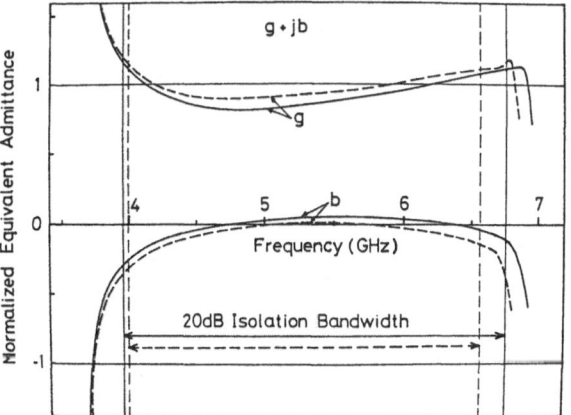

Fig. 8.22. Normalized equivalent admittance of the circulator whose performance is shown in Fig. 8.21 [8.2]

(solid curves) are close to their ideal values (1 and 0) over bandwidths wider than in the best triangular circulator (broken curves) and the best disk-shaped circulator (Fig. 8.16).

Miyoshi and *Miyauchi* gave scaling rules to generalize the above design concept so that it can be applied to different frequencies and different ferrite materials [8.2].

8.6 Summary

The analysis and design of a planar circuit having anisotropic spacing material have been presented, and their application to the analysis and optimum design of a triplate-type stripline circulator has been described. It has been shown that the planar circuit concept is a powerful tool in the design of such a device to which the one-dimensional, distributed-constant line model does not apply.

9. Optical Planar Circuits

The final two chapters are devoted to the theory of the optical planar circuit which is defined as *a circuit having dimensions comparable to the wavelength in one direction but much greater than it in the other two directions.* The technical significance of this circuit concept has increased throughout the 1970s because it plays an important role in some types of optical integrated circuitry.

In this chapter, basic theories of optical planar circuits based on wave optics and geometrical optics are presented first. In the latter half, waves in a uniform optical planar circuit, optical planar circuits having slowly varying thickness, and those having periodic structures and planar lenses are considered.

9.1 Background

The so-called optical IC (integrated circuits) can basically be classified into two groups: the waveguide type, in which light energy is confined in linear and curved channels, and the planar type, in which it is confined in a two-dimensional space. The theory of the optical planar circuit is indispensable in designing the second group of optical ICs, and also in designing some of the first group (Chapt. 10). The historical and technological significance of the optical planar circuit concept has been discussed in the second paragraph of Sect. 1.1.4. Readers are expected to review this short paragraph before proceeding to the following description.

We consider a dielectric two-dimensional waveguiding structure, as shown in Fig. 9.1. The sandwiched waveguiding region (I) and the upper and lower

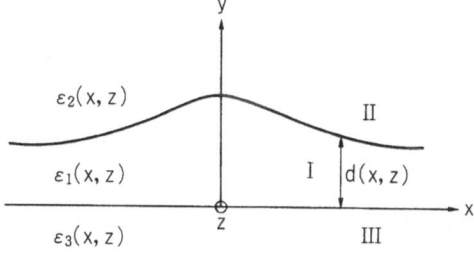

Fig. 9.1. An asymmetrical dielectric two-dimensional waveguiding structure

substrate regions (II and III) have permittivities of ε_1, ε_2, and ε_3, respectively. These are generally considered to be functions of position (x, z); however, in many cases in the following discussions these are assumed to be constant. The thickness of the waveguiding region is assumed to be given as a function of position.

The optical planar circuit is essentially different from the planar circuit in microwave or millimeter-wave regions in several aspects:

1) As to the circuit-analysis technique, the mode-expansion method and contour-integral method are commonly used in microwave/millimeter-wave regions. However, these methods are not practical for the optical planar circuit because it has dimensions much greater than the wavelength in two dimensions, and hence requires an impractically long computer time.
2) Because metals are dissipative at optical frequencies, in optical planar circuits the substrate must be dielectrics such as glass, semiconductors, or polymer materials. These dielectrics have refractive indices ranging between $1.4-3.0$, and are not able to constitute a definite boundary between the waveguiding and substrate regions. Thus *leaky wave* phenomena often make the circuit performance complex.
3) A spatially varying optical planar circuit in which the effective refractive index is given as a function of position is not difficult to fabricate and is widely used, whereas its microwave/millimeter-wave counterpart is scarcely used.

As a consequence of the above differences, the development of a new system of theories for the analysis and design of the optical planar circuit is needed. Such theories, however, have neither been developed satisfactorily nor well systematized. Our present knowledge can be classified into the following four items:

1) A wave-optics approach based on the concept of the *effective refractive index* (Sect. 9.2).
2) A geometrical optics approach based on a two-dimensional eikonal equation. This approach is applicable to cases where the spatial variation of the effective refractive index is gradual as compared with wavelength, and is often used in the analysis and design of optical planar circuit components such as geodesic lenses (Sect. 9.6).
3) A wave-optics approach based on the concept of the *optical Bloch wave*. This is a somewhat special technique developed for analyzing optical planar circuits having periodic structures (Sect. 9.5).
4) The beam-propagation method (BPM) for analyzing optical planar circuits having stripelike waveguiding structures.

In this chapter, the first three items are described in as systematic a manner as possible. The fourth item is dealt with separately in the next chapter.

9.2 Wave-Optics Approach to Optical Planar Circuits

9.2.1 Basic Equations

We assume that in a structure like the one shown in Fig. 9.1, the waveguide thickness d and permittivities ε_1, ε_2, and ε_3 are all functions of position (x, z), but the variations are gradual as compared with the wavelength of light. We further assume that $\varepsilon_1 > \varepsilon_2 > \varepsilon_3$, and all the materials are nonmagnetic. In such a case, the electric and magnetic fields in each region, E_i and H_i ($i = 1, 2, 3$), are governed by the following differential equations:

$$\nabla^2 E_i + \omega^2 \varepsilon_i \mu_0 E_i + (\nabla \ln \varepsilon_i \cdot \nabla) E_i$$

$$+ (E_i \cdot \nabla) \nabla \ln \varepsilon_i + \nabla \ln \varepsilon_i \times (\nabla \times E_i) = 0 , \tag{9.1}$$

$$\nabla^2 H_i + \omega^2 \varepsilon_i \mu_0 H_i + \nabla \ln \varepsilon_i \times (\nabla \times H_i) = 0 . \tag{9.2}$$

If we approximate $\nabla \ln \varepsilon_i \simeq 0$ in the above equations because variations of ε_i are assumed to be gradual, simplified expressions for E_{yi} and H_{yi} are obtained from the above equations [9.1]:

$$[\nabla^2 + \omega^2 \varepsilon_i(x, z) \mu_0] E_{yi} = 0 , \tag{9.3}$$

$$[\nabla^2 + \omega^2 \varepsilon_i(x, z) \mu_0] H_{yi} = 0 . \tag{9.4}$$

Other field components can be expressed in terms of E_{yi} and H_{yi}. These expressions are derived directly from Maxwell's equations as

$$\left(\omega^2 \varepsilon \mu_0 + \frac{\partial^2}{\partial y^2} \right) E_x = \frac{\partial^2}{\partial x \, \partial y} E_y + j \omega \mu_0 \frac{\partial}{\partial z} H_y , \tag{9.5a}$$

$$\left(\omega^2 \varepsilon \mu_0 + \frac{\partial^2}{\partial y^2} \right) E_z = \frac{\partial^2}{\partial z \, \partial y} E_y - j \omega \mu_0 \frac{\partial}{\partial x} H_y , \tag{9.5b}$$

$$\left(\omega^2 \varepsilon \mu_0 + \frac{\partial^2}{\partial y^2} \right) H_x = -j \omega \varepsilon \frac{\partial}{\partial z} E_y + \frac{\partial^2}{\partial x \, \partial y} H_y , \tag{9.5c}$$

$$\left(\omega^2 \varepsilon \mu_0 + \frac{\partial^2}{\partial y^2} \right) H_z = j \omega \varepsilon \frac{\partial}{\partial x} E_y + \frac{\partial^2}{\partial z \, \partial y} H_y , \tag{9.5d}$$

where subscript i is omitted for simplicity.

In an optical planar circuit, we have two kinds of modes which are linearly independent of each other: one with $E_{yi} = 0$ everywhere and the other with $H_{yi} = 0$ everywhere. We next consider these modes separately [9.2].

9.2.2 TE Modes

A mode in which $E_{yi} = 0$ everywhere is called a TE (transverse electric) mode, because when such a mode is propagated in the z direction uniformly with respect to the x axis (i.e., $\partial/\partial x = 0$), $E_{zi} = 0$ holds everywhere (9.5b).

Let H_{yi} be expressed as

$$H_{yi} = Y_i(y \mid \varepsilon_i(x, z), d(x, z)) \cdot X(x, z) , \tag{9.6}$$

and put into (9.4). If we assume that $\partial Y_i/\partial x \simeq 0$ and $\partial Y_i/\partial z \simeq 0$, we obtain

$$\frac{\frac{\partial^2}{\partial y^2} Y_i}{Y_i} + \frac{\left(\frac{\partial^2}{\partial x^2} + \frac{\partial^2}{\partial z^2}\right) X + k_0^2 n_i^2(x, z) X}{X} = 0 , \tag{9.7}$$

where $k_0 = \omega\sqrt{\varepsilon_0\mu_0}$ and $n_i(x, z)$ denote the refractive index of the ith region $[n_i = (\varepsilon_i/\varepsilon_0)^{1/2}]$. We note here that the two terms in the left-hand side are mutually independent, and equate these to $\gamma_i^2(x, z)$ and $-\gamma_i^2(x, z)$, respectively. Thus we find that Y_i and X satisfy separately the following equations:

$$\frac{\partial^2}{\partial y^2} Y_i = \gamma_i^2(x, z) Y_i , \tag{9.8}$$

$$\left(\frac{\partial^2}{\partial x^2} + \frac{\partial^2}{\partial z^2}\right) X + k_0^2 N^2(x, z) X = 0 , \tag{9.9}$$

where

$$k_0^2 N^2(x, z) = k_0^2 n_i^2(x, z) + \gamma_i^2(x, z) . \tag{9.10}$$

The above equations tell us that the separation parameter γ_i directly gives the vertical (y-direction) wavenumber, whereas $N(x, y)$ determines the two-dimensional wave behavior.

Since the light wave is confined in the region 1, we may postulate that γ_1 is imaginary, whereas γ_2 and γ_3 are real. Therefore, to simplify the equations, we write

$$\gamma_1 = j\kappa = jk_0\sqrt{n_1^2 - N^2} , \tag{9.11a}$$

$$\gamma_2 = p = k_0\sqrt{N^2 - n_2^2} , \tag{9.11b}$$

$$\gamma_3 = q = k_0\sqrt{N^2 - n_3^2} , \tag{9.11c}$$

and henceforth use κ, p, and q instead of γ_i's.

We first solve (9.8). The solution of this equation is given as an exponential or a sinusoidal function. From the conditions that Y_2 and Y_3 must be zero at $y = \pm \infty$, we obtain

$$Y_2 = A\,e^{-p(y-d)}, \tag{9.12a}$$

$$Y_1 = B\cos(\kappa y + \phi_1), \tag{9.12b}$$

$$Y_3 = C\,e^{qy}. \tag{9.12c}$$

Parameters p, κ, q can be determined by the boundary conditions. From the conditions that E_x, E_z, H_x, and H_z must all be continuous at the lower boundary ($y = 0$) and the upper boundary ($y = d$), we find that Y and ($\partial Y/\partial y$) must also be continuous [see (9.5)]. Hence, from (9.12), p, κ, and q must satisfy a relation

$$\kappa d = \tan^{-1}(p/\kappa) + \tan^{-1}(q/\kappa) + m\pi \quad (m = 0, 1, 2, \ldots), \tag{9.13}$$

where m denotes the mode number. We may determine p, κ, q, and also N by solving (9.11a – c, 13) simultaneously, for each TE mode.

The parameter N thus determined is called the *effective refractive index* because an optical planar circuit having any distribution of ε_1, ε_2, ε_3, and d can simply be modeled as a two-dimensional waveguide having the index distribution of $N(x, z)$. Thus the analysis of an optical planar circuit is attributed to the problem of solving a two-dimensional wave equation given as (9.9). Note that $N(x, z)$ is defined for each TE mode, and satisfies $\varepsilon_3 < N < \varepsilon_1$ when the mode is a propagating mode (9.11a – c).

The x and z components of the fields in TE modes are given, from (9.5a – d), as

$$E_x = \frac{j\omega\mu_0}{k_0^2 N^2}\, Y\, \frac{\partial X}{\partial z}, \tag{9.14a}$$

$$E_z = -\frac{j\omega\mu_0}{k_0^2 N^2}\, Y\, \frac{\partial X}{\partial x}, \tag{9.14b}$$

$$H_x = \frac{1}{k_0^2 N^2}\, \frac{\partial Y}{\partial y}\, \frac{\partial X}{\partial x}, \tag{9.14c}$$

$$H_z = \frac{1}{k_0^2 N^2}\, \frac{\partial Y}{\partial y}\, \frac{\partial X}{\partial z}. \tag{9.14d}$$

9.2.3 TM Modes

A mode in which $H_{yi} = 0$ everywhere is called a TM (transverse magnetic) mode, because when such a mode is propagated in the z direction uniformly with respect to the x axis (i.e., $\partial/\partial x = 0$), $H_{zi} = 0$ holds everywhere (9.5d).

Following the preceding calculation of the TE modes, we express E_{yi} as

$$E_{yi} = Y_i(y\,|\,\varepsilon_i(x, z), d(x, z)) \cdot X(x, z). \tag{9.15}$$

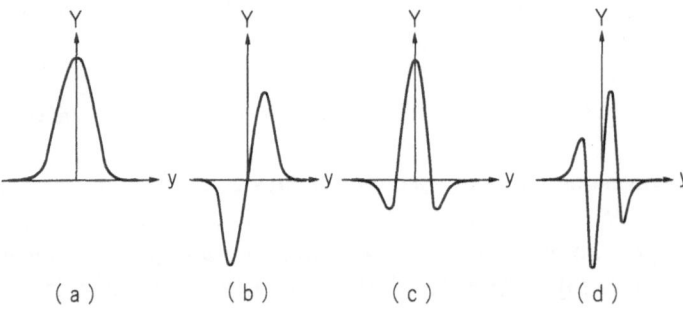

Fig. 9.2a – d. Vertical distributions of the field function $Y(y)$ for four lowest-order modes: **(a)** $m = 0$, **(b)** $m = 1$, **(c)** $m = 2$, **(d)** $m = 3$

Then, after calculations similar to those for the TE modes, we obtain Y_2, Y_1, and Y_3 identical to (9.12a, b, c), respectively, and

$$\kappa d = \tan^{-1} \frac{p}{\kappa} \frac{\varepsilon_1}{\varepsilon_2} + \tan^{-1} \frac{q}{\kappa} \frac{\varepsilon_1}{\varepsilon_3} + m\pi \quad (m = 0, 1, 2, \ldots), \tag{9.16}$$

$$E_x = \frac{1}{k_0^2 N^2} \frac{\partial Y}{\partial y} \frac{\partial X}{\partial x}, \tag{9.17a}$$

$$E_z = \frac{1}{k_0^2 N_2} \frac{\partial Y}{\partial y} \frac{\partial X}{\partial z}, \tag{9.17b}$$

$$H_x = -\frac{j\omega\varepsilon}{k_0^2 N^2} Y \frac{\partial X}{\partial z}, \tag{9.17c}$$

$$H_z = \frac{j\omega\varepsilon}{k_0^2 N^2} Y \frac{\partial X}{\partial x}, \tag{9.17d}$$

where m denotes again the mode number.

9.2.4 Vertical Field Distributions

The vertical distributions of the field function Y are computed for the first four modes, $m = 0$, 1, 2, 3, and are shown in Fig. 9.2. Such curves are common in both the TE and TM modes. Numerical calculation of (9.13, 16) shows that the fundamental mode (the mode having lowest cutoff frequency and highest propagation constant) is the TM_0 mode, where the subscript 0 denotes $m = 0$ [9.2].

9.3 Geometrical Optics Approach to Optical Planar Circuits

9.3.1 Concepts of Geometrical Optics and Ray

An important step in the analysis of an optical planar circuit is to understand the behavior of light beams in a spatially inhomogeneous medium, in which the refractive index is expressed generally as a function of position. Two approaches are known for such problems. The first is to solve Maxwell's equations or the wave equation to obtain solutions expressing the wave behavior; this approach has been described in Sect. 9.2. The second is to investigate the path of a ray of light traveling through the medium (geometrical optics).

The ray concept is applicable only to those cases in which the beam diameter is much larger than the wavelength of light. A beam having a diameter comparable to the wavelength can never be treated as a ray, because such a beam is subject to an immediate divergence due to diffraction (Sect. 9.4.2), which is a phenomenon beyond geometrical optics. On the other hand, if the beam diameter is much larger than the wavelength of light, an approximate wave equation expressing the ray path can be derived from Maxwell's equations. The real scalar variable used in such an equation to express the wavefront phase is called the *eikonal*, and the equation is called the *eikonal equation*.

9.3.2 Eikonal and Eikonal Equation

A mathematically simple derivation of the two-dimensional eikonal equation follows. (For a more comprehensive, but mathematically less simple derivation see Ref. [9.1], Sects. 2.7, 8.)

We consider a case where the effective refractive index $N(x, z)$ is only slowly varying with respect to the position (x, z), more strictly, when its variation over one wavelength of light is negligible. We assume that the solution of (9.9) can be expressed as

$$X(x, z) = K(x, z) \exp[-jk_0 \phi(x, z)] , \tag{9.18}$$

where $K(x, z)$ is a real amplitude function, and substitute this equation into (9.9). Then we obtain a differential equation governing $K(x, z)$:

$$k_0^2(N^2 - |\nabla \phi|^2)K - jk_0(2\nabla\phi\nabla K + K\nabla^2\phi) + \nabla^2 K = 0 , \tag{9.19}$$

where

$$\nabla = i_x(\partial/\partial x) + i_z(\partial/\partial z) , \tag{9.20}$$

and i_x and i_z denote unit vectors in the x and z directions, respectively. The function $\phi(x, z)$ is a scalar real function expressing the phase at position (x, z); this function is called the eikonal.

If we take the above slowly varying assumption into consideration, we see that the third term can be neglected as compared with the first term. Furthermore, because k_0, ϕ, and K are all real, the first term is real whereas the second is imaginary. Therefore we obtain, separating the real part,

$$k_0^2(N^2 - |\nabla\phi|^2)K(x, z) = 0, \tag{9.21}$$

which leads directly to $|\nabla\phi|^2 = N^2$, i.e.,

$$\left(\frac{\partial\phi}{\partial x}\right)^2 + \left(\frac{\partial\phi}{\partial y}\right)^2 = N^2(x, z). \tag{9.22}$$

This is the equation that governs the behavior of eikonal ϕ. As shown in the next section, an equieikonal surface on which

$$\phi(r) = \text{const} \tag{9.23}$$

corresponds to a wavefront of light, and the ray direction is given by a curve normally intersecting the equieikonal surfaces. Equation (9.22) is called the *eikonal equation*; this is the basic equation for the geometrical-optics solution in inhomogeneous media.

9.3.3 Ray Equation

From the eikonal equation we derive an equation which is called the ray equation. For the time being, we consider the problem in a three-dimensional space, and note that the power-flow density along a light beam is given as the real part of the complex Poynting vector defined as

$$S = \text{Re}\{\tfrac{1}{2}E \times H^*\}, \tag{9.24}$$

where E and H denote electric and magnetic fields. On the other hand, E and H can be expressed in forms similar to (9.18) as

$$E = e(r)\exp[-jk_0\phi(r)], \tag{9.25a}$$

$$H = h(r)\exp[-jk_0\phi(r)], \tag{9.25b}$$

where r is a three-dimensional position vector.

Using e and h, we may rewrite (9.24) as

$$S = \text{Re}\{\tfrac{1}{2}e \times h^*\}, \tag{9.26}$$

and rewrite it further, after algebraic computations described in Appendix A9.1, as

$$S = (1/2c\mu)\,\mathrm{Re}\{(e \cdot e^*)\,\nabla\phi\} = (1/2c\mu)(e \cdot e^*)\nabla\phi\,. \tag{9.27}$$

Since $(e \cdot e^*)$ in the above equation is a real scalar quantity corresponding to the mean stored energy of the electric field, this equation tells us that the energy flow direction, i.e., the ray direction, coincides with that of $\nabla\phi$. Further, from the three-dimensional eikonal equation, $|\nabla\phi|^2 = n^2$, where ∇ denotes a three-dimensional vector operator,

$$|\nabla\phi| = n\,. \tag{9.28}$$

Hence, the unit vector s along the ray is expressed as

$$s = \nabla\phi/|\nabla\phi| = \nabla\phi/n\,. \tag{9.29}$$

An alternative expression for the ray is obtained by using a curved coordinate s along the ray. Since

$$dr/ds = s\,, \tag{9.30}$$

(9.29) can be rewritten as

$$n(x, y, z)\,dr/ds = \nabla\phi\,. \tag{9.31}$$

Further, after some calculations described in Appendix A 9.2, (9.31) can be rewritten as

$$(d/ds)[n(x, y, z)s] = \nabla n\,, \quad \text{or} \tag{9.32}$$

$$(d/ds)[n(x, y, z)\,dr/ds] = \nabla n\,. \tag{9.33}$$

Equation (9.32) or (9.33) is the basic equation determining the ray path in an inhomogeneous medium. These equations are called ray equations.

We now go back to a two-dimensional, optical planar circuit problem. In this case, because the circuit can be replaced by a two-dimensional model (uniform in the y direction) having an effective index of $N(x, z)$, we may rewrite (9.32, 33) as

$$\frac{d}{ds}[N(x, z)s] = \nabla N\,, \tag{9.34}$$

$$\frac{d}{ds}\left[N(x, z)\frac{dr}{ds}\right] = \nabla N\,, \tag{9.35}$$

where s, s, and r are now all defined in a two-dimensional space and ∇ denotes again a two-dimensional vector operator (9.20).

We should note that (9.34, 35) are applicable only to the two-dimensional equivalent model of a planar circuit. In an actual planar circuit, the ray is sometimes considered to travel in Region 1 (Fig. 9.1) along a zigzag path being reflected at the boundaries between Region 1 and Regions 2 or 3 [9.3, 4]. When the discussion is based upon such a model, we have to use the three-dimensional ray equation (9.32) or (9.33) rather than (9.34) or (9.35).

9.4 Optical Planar Circuits Having Uniform Slab Structure

9.4.1 Model to be Considered

We consider the simplest structure in which ε_1, ε_2, ε_3 are all constants (Fig. 9.1), and moreover d is constant everywhere. In this case, the effective index N is also constant everywhere.

In such a structure, several simple mathematical solutions of the wave equation are obtained; these are the plane wave, which is uniform in the x direction when propagating in the z direction; the two-dimensional Hermite-Gaussian wave; and diverging and converging cylindrical waves. In the following we discuss first the TE-mode waves and later the TM-mode ones.

9.4.2 TE-Mode Waves

a) Plane Wave

We consider a TE-mode wave ($E_y = 0$ everywhere) uniform in the x direction and propagating in the z direction. If we write the solution of (9.9) as

$$X(x, z) = X_0 e^{-j\beta z}, \tag{9.36}$$

where X_0 is a constant, we have

$$\beta^2 = k_0^2 N^2 = k_0^2 n_1^2 - \kappa^2. \tag{9.37}$$

Therefore, the eigenequation (dispersion relation) of this planar circuit is given from (9.13) as

$$\tan(\kappa d) = \frac{\kappa(p+q)}{\kappa^2 - pq}. \tag{9.38}$$

Furthermore, (9.14b, 36) lead to $E_z = 0$, which means that this is truly a TE mode.

b) Hermite-Gaussian Wave

We consider a wave propagating approximately in the z direction and write [9.5]

$$X(x, z) = X_1(x, z) e^{-j\beta z} .$$

(9.39)

If we assume here concerning X_1 that

$$2\beta \left| \frac{\partial X_1}{\partial z} \right| \gg \left| \frac{\partial^2 X_1}{\partial z^2} \right| ,$$

(9.40)

we find, substituting (9.39) into (9.9), that X_1 should satisfy

$$\frac{\partial^2 X_1}{\partial x^2} - 2j\beta \frac{\partial X_1}{\partial z} + (k^2 N^2 - \beta^2) X_1 = 0 .$$

(9.41)

The above equation has a special solution nonuniform both in the x and z directions, which is expressed as

$$X_1(x, z) = H_n\left(\frac{x}{W(z)}\right) \exp\left[-\frac{1}{2}\left(\frac{x}{W(z)}\right)^2\right] \exp\left(-j\beta \frac{x^2}{2R(z)}\right)$$

$$\times \exp[j(n+1)\phi(z)] \exp(-j\beta z) ,$$

(9.42)

where H_n denotes the nth order Hermite polynomial, and functions $W(z)$, $R(z)$, and $\phi(z)$ are defined as

$$W(z) = W_0[1 + (2z/\beta W_0^2)^2]^{1/2} ,$$

(9.43)

$$R(z) = z[1 + (\beta W_0^2/2z)^2] ,$$

(9.44)

$$\phi(z) = \tan^{-1}(2z/\beta W_0^2) .$$

(9.45)

Waves expressed by (9.42) are called Hermite-Gaussian waves.

To investigate the property of these waves, we note first that the amplitude of X_1 is given by the first and second factors — i.e., H_n and the Gaussian function: $\exp[-(1/2)(x/W)^2]$ — whereas the following three factors contribute only to the phase variation. The Hermite polynomials are pseudo-periodic functions, whereas the Gaussian function is a symmetrical, rapidly decreasing function. Therefore, the initial amplitude variations of the nth order Hermite-Gaussian waves, $X_{1n}(x, 0)$, are as shown in Fig. 9.3a.

Furthermore, the arguments of both the Hermite and Gaussian factors are given as $x/W(z)$, where $W(t)$ is given by (9.43). This equation expresses a function increasing in a hyperbolic manner, as shown in Fig. 9.3b. Combining Figs. 9.3a, b, we find that a Hermite-Gaussian wave expresses an expanding two-dimensional light beam in which the shape of the lateral amplitude distribution is preserved.

We should note that in such a wave both E_z and H_z exist. Such a mode is often called a hybrid mode in waveguide theory. The amplitude of E_z, however, is relatively small in the present case.

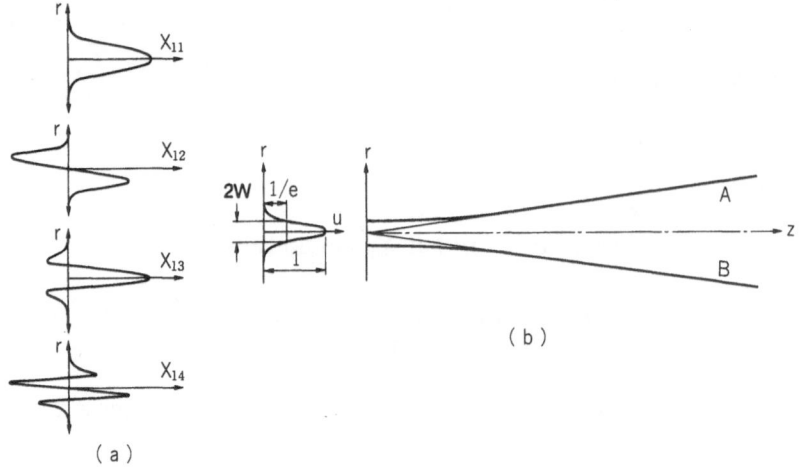

Fig. 9.3. (a) The transverse amplitude distribution of the nth order Hermite-Gaussian mode. **(b)** Spread of a light beam of the Hermite-Gaussian mode

c) Cylindrical Waves

If we substitute $x = r \cos \phi$ and $z = r \sin \phi$ into (9.9), we find that so-called cylindrical waves exist, which are axially symmetrical with respect to the y axis. These waves are expressed as

$$X_1(r, \phi) = H_n^{(1)}(\beta r) \cos(n\phi + \theta_n) e^{j\beta r}, \quad \text{or} \tag{9.46a}$$

$$X_2(r, \phi) = H_n^{(2)}(\beta r) \cos(n\phi + \theta_n') e^{-j\beta r}, \tag{9.46b}$$

where $\beta = k_0 N_0$, and $H_n^{(1)}$ and $H_n^{(2)}$ are the nth order Hankel functions of the first and second kinds, respectively. These are defined as

$$H_n^{(1)}(\beta x) = J_n(\beta x) + j N_n(\beta x), \tag{9.47a}$$

$$H_n^{(2)}(\beta x) = J_n(\beta x) - j N_n(\beta x), \tag{9.47b}$$

where J_n and N_n denote the nth order Bessel and modified Bessel functions, respectively.

Equations (9.46a, b) express radially converging and radially diverging cylindrical waves, respectively. Both of these wave have no radial electric field component.

9.4.3 TM-Mode Waves

In this case, the plane wave, Hermite-Gaussian waves, and cylindrical waves can be derived in a manner similar to the TE-mode case. Most of the equations are identical; however, (9.38) must be rewritten as

$$\tan(\kappa d) = \frac{\kappa \left(p \dfrac{\varepsilon_1}{\varepsilon_2} + q \dfrac{\varepsilon_1}{\varepsilon_3} \right)}{\kappa^2 - pq \dfrac{\varepsilon_1^2}{\varepsilon_2 \varepsilon_3}} . \tag{9.48}$$

9.5 Optical Planar Circuits Having Periodic Structures

We next consider optical planar circuits which have one-dimensional or two-dimensional periodic structures. Since a wave existing in such a circuit exhibits properties the same as those of a Bloch wave (the electron wave in a periodic crystalline structure), it is often called an "optical Bloch wave." In this section the behavior of the optical Bloch waves and their applications are considered following *Ulrich* and *Zengerle* [9.6].

9.5.1 Mathematical Expression of Optical Bloch Waves

We consider an optical planar circuit having an one-dimensional or two-dimensional periodic structure, for example, the periodic variation of the slab thickness d.

The simplest example to begin with is a one-dimensional periodic structure in the z direction having a period L. In such a structure, a traveling wave $\exp(-j\beta_0 z + j\omega t)$ is "spatially modulated" by a stationary electromagnetic field distribution. Hence, any one of the six electromagnetic field components can be expressed generally as

$$A(x, y, z, t) = \sum_{n=0}^{\infty} A_n \cos\left(2\pi n \frac{z}{L}\right) \exp(\mp j\beta_0 z + j\omega t)$$

$$= \sum_{n=-\infty}^{\infty} A_n \exp(-j\beta_n z + j\omega t), \tag{9.49}$$

where

$$\beta_n = \pm \beta_0 k + nK \quad (n = \ldots, -2, -1, 0, 1, 2, \ldots). \tag{9.50}$$

Here k denotes a unit vector in z direction, and K is a vector called the "reciprocal lattice vector," which is defined as

$$K = \{0, 0, 2\pi/L\}. \tag{9.51}$$

In (9.49, 50), the upper and lower signs give the forward and backward waves, respectively.

When the periodic structure is two-dimensional, in other words, when periodicity is present in both the x and z directions, (9.49, 50) should be re-written as

$$A(x, y, z, t) = \sum_{m=-\infty}^{\infty} \sum_{n=-\infty}^{\infty} A_{mn} \exp(j\beta_{nm}r + j\omega t) \tag{9.52}$$

and

$$\beta_{mn} = \beta_0 + mK_x + nK_z. \tag{9.53}$$

Equations (9.49, 52) show that the optical wave in a periodic structure can be expressed as a superposition of many *spatial harmonic* waves. The fact that such a spatial harmonic expression is possible is called Floquet's theorem.

9.5.2 Dispersion Relation and Group Velocity of Optical Bloch Waves

The dispersion relation of an optical Bloch wave is determined from the rela-tion between the angular frequency ω and the propagation constant β. We consider first a plane wave propagating in the z direction, in a one-dimen-sional periodic structure also in the z direction. In such a case, the function $\omega(\beta_n)$ can be expressed by periodic curves having a common periodicity K_z, as shown in Fig. 9.4. In this figure, the slant straight lines show the relations given by (9.50), whereas the actual dispersion curves "jump" between these straight lines smoothly because of the electromagnetic coupling between space harmonic modes. We note in Fig. 9.4 that the magnitude of the group velocity, which is given by $|d\omega/d\beta|$, is determined solely by the frequency ω and does not depend on the mode number n. Note also, however, that forward waves (for which $d\omega/d\beta > 0$) and backward waves (for which $d\omega/d\beta < 0$) exist.

Next we consider the case when a plane wave is propagated in any direc-tion in the x-z plane, whereas the periodic structure is present again only in the z direction. In such a case, one method of the graphic expression is to plot the propagation constants of a wave in a $\beta_z - \beta_x$ plane, as shown in Fig. 9.5, as-

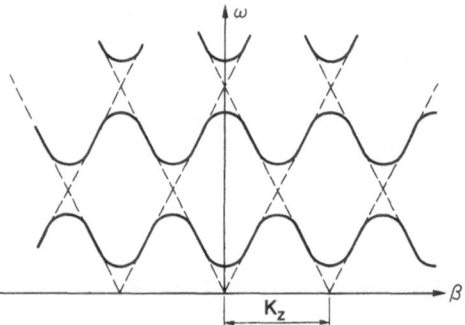

Fig. 9.4. The dispersion diagram of an optical planar circuit having a periodic structure

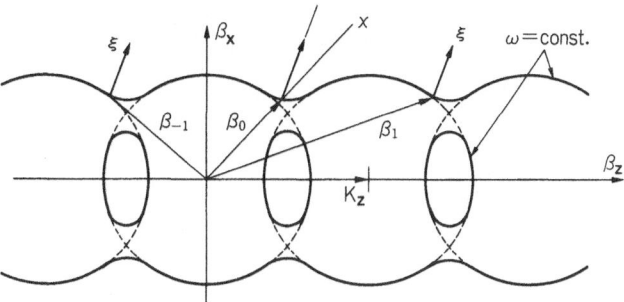

Fig. 9.5. The contour of $\omega = $ const plotted on a $\beta_z - \beta_x$ diagram [9.6]

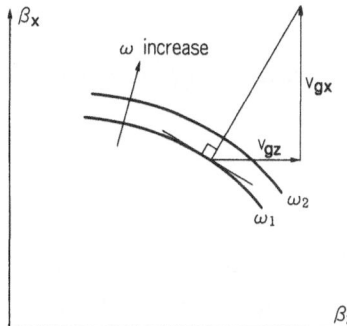

Fig. 9.6. A diagram illustrating the direction of the group-velocity vector

suming that $\omega = $ const. As seen in Fig. 9.5, the fundamental mode is expressed by a circle given by $(\beta_z^2 + \beta_x^2)^{1/2} = $ const, whereas the space harmonics are expressed by the circles translated in the z direction by nK_z where n is an integer. Furthermore, because of the electromagnetic coupling existing between modes, the actual dispersion curves become like the thick solid curves in Fig. 9.5.

A comment should be added regarding the directions of the phase-velocity vector and the group-velocity vector. The direction of the phase-velocity vector naturally coincides with that of

$$\boldsymbol{\beta}_r = \boldsymbol{i}\,\beta_x + \boldsymbol{k}\,\beta_z \,, \tag{9.54}$$

where $\boldsymbol{i}$ and $\boldsymbol{k}$ are unit vectors in the x and z directions, respectively. On the other hand, the direction of the group-velocity vector is normal to the equi-ω contour, and headed along the increase of ω, as shown in Fig. 9.6. The above statement will be readily understood if we notice that the group-velocity components are given as

$$v_{gx} = d\omega/d\beta_x \,, \tag{9.55a}$$

$$v_{gz} = d\omega/d\beta_z \,, \tag{9.55b}$$

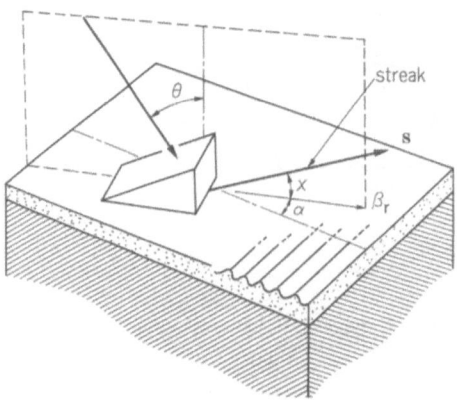

Fig. 9.7. Excitation of Bloch waves by a prism coupler [9.6]

and that two infinitesimally close equi-ω contours are parallel with each other as shown in Fig. 9.6.

9.5.3 Excitation of Optical Bloch Waves

The simplest method for exciting a specific Bloch wave is to use a prism coupler, as shown in Fig. 9.7. We first determine the angle α shown in Fig. 9.7 (the direction of the phase-velocity vector) and the polarization of the wave to be excited, that is, whether the TE mode or TM mode should be excited. We place the prism coupler in the direction of α, and adjust the incident angle θ until a strong streak is found, in Fig. 9.7 in the direction of vector s. This means that at this incident angle θ the phase match is achieved.

The direction of s is nothing but the direction of the group-velocity vector. A significant feature of Bloch waves is that the phase-velocity and group-velocity vectors have generally different directions. In some cases, these two vectors become almost normal to each other. The angle χ in Fig. 9.7 is called the *beam steering angle* in the theory of surface acoustic waves [9.7].

9.5.4 Example of Measured $\beta_z - \beta_x$ Relations

From the values of α and θ, we can compute (β_x/β_z) and $(\beta_z^2 + \beta_x^2)$, respectively. Hence, using the measured relation of α versus θ, we can plot the measured characteristics on a $\beta_z - \beta_x$ diagram.

An example of such plotting (a small area around a crosspoint of two circular loci in Fig. 9.5) is shown in Fig. 9.8. In this case, the periodic optical planar circuit consists of a 165 nm-thick Ta_2O_5 layer (refractive index: 2.10) sputtered on a glass substrate (refractive index: 1.47), and a 48 nm-deep sinusoidal grating with a spatial period $L_z = 282.2$ nm. The wavelength of the light used in the experiment is 611.7 nm. Because both the TE and TM modes exist, all

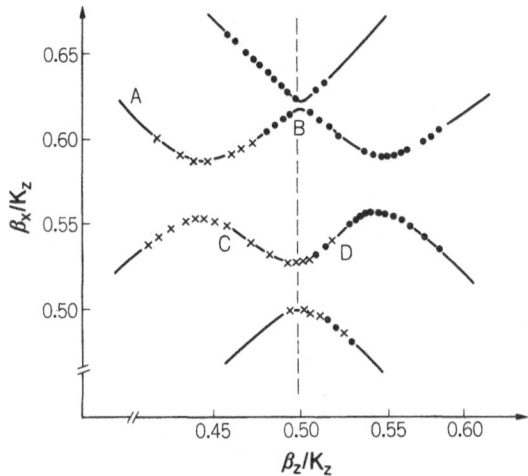

Fig. 9.8. Measured dispersion contours on a $\beta_z - \beta_x$ diagram. Dots and small crosses show results obtained with TE and TM excitations, respectively. Angular frequency $\omega = 3.08 \times 10^{15}\,\mathrm{s}^{-1}$ [9.6]

of the circular loci in Fig. 9.5 become double circles, thus making the loci in Fig. 9.8 rather complex. The coupling constants K between modes can be calculated from the *gap* $\Delta\beta_x$ between the coupled mode curves as

$$K = L_z \Delta\beta_x / 4\pi\,, \tag{9.56}$$

which gives in the present case

$$K_{\text{TE-TE}} = 0.003\,, \quad K_{\text{TE-TM}} = 0.017\,, \quad K_{\text{TM-TM}} = 0.012\,. \tag{9.57}$$

9.6 Planar Lenses

In the design of optical planar ICs, the planar lens is one of the most important components.

The principle for the design of some planar lenses is essentially identical to that for conventional lenses, whereas it is entirely different for other types. Figure 9.9 shows three basic types of the planar lens [9.8]. These are

1) the "mode-index lens," which uses the spatial variation of the effective refractive index defined in Sect. 9.2.2 (Fig. 9.9a);
2) the "geodesic lens," which uses the transmission of light rays along a "geodesic" in a curved planar light-guiding structure as shown in Fig. 9.9b [note that in such a structure a ray travels along a geodesic (shortest path) according to Fermat's principle];
3) the "grating lens," which uses Bragg diffraction of light waves in a planar periodic structure (Fig. 9.9c).

The principle of the index lens is self-explanatory. Actually, there are some versions in this type of planar lens, which are shown in Fig. 9.10 [9.9]. These

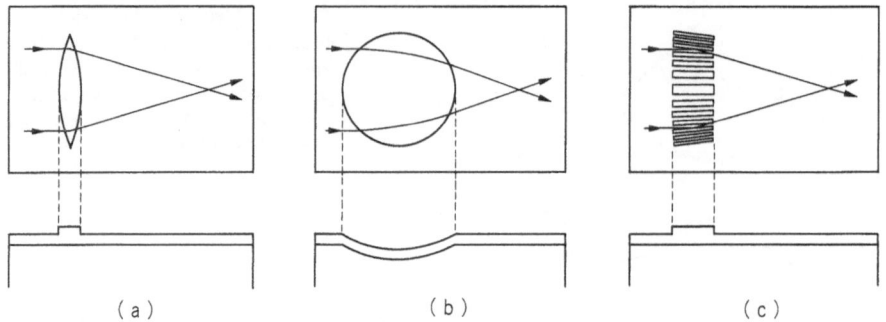

Fig. 9.9a – c. Three basic types of the planar lens: **(a)** mode-index lens, **(b)** geodesic lens, **(c)** grating lens [9.8]

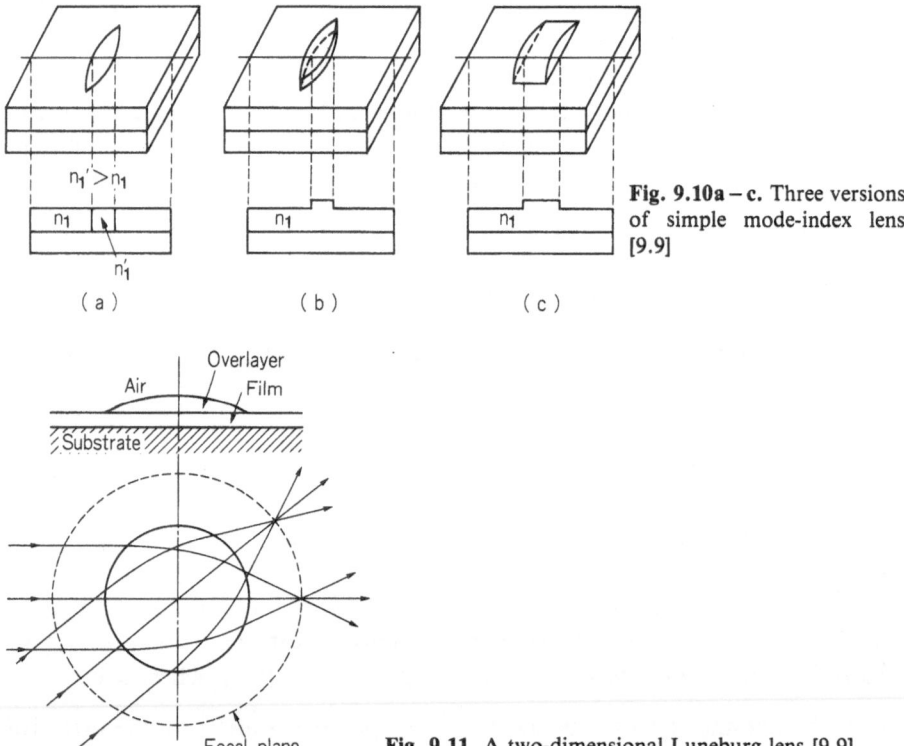

Fig. 9.10a – c. Three versions of simple mode-index lens [9.9]

Fig. 9.11. A two-dimensional Luneburg lens [9.9]

lenses correspond to the so-called single lenses in conventional optics, and hence have rather large aberrations. An aberration-free focus can be obtained with a *Luneburg* lens [9.10, 11] which was used first at microwave frequencies in high-speed beam scanners. The structure of a two-dimensional Luneburg lens is shown in Fig. 9.11 [9.9], which shows an axially symmetrically thickened optical planar circuit. The thickness is controlled so that an aberration-free focus is obtained along the curved focal plane shown in the figure.

The mode-index lenses mentioned in the preceding paragraph can all be analyzed and designed by the principles described in this chapter. The geodesic and grating lenses, however, need slightly different approaches, which are omitted here because of space limitations.

9.7 Summary

Basic theories of optical planar circuits based on the wave-optics and geometrical optics approaches have been presented, and applied to various types of optical waves in uniform and periodic optical planar circuits. Generally speaking, this academic area has not been fully exploited; further progress in theoretical tools is still expected.

A specific category of the optical planar circuit is those having a stripe structure, in other words, those in which the light energy is guided along a guiding structure. This category of circuits will be dealt with separately in the following chapter.

10. Optical Planar Circuits Having Stripelike Waveguiding Structures

This brief final chapter is the second of the two chapters describing the theory of optical planar circuits. The objective of the analysis is somewhat special in this chapter: optical planar circuits having stripelike waveguiding structures. The analysis methods have not been well developed in this area. Only the beam-propagation method (BPM) is described as an approach particularly suited to the analysis of such circuits. The light-guiding behavior of a tapered waveguide and a Y-junction circuit are calculated as examples.

10.1 Background

As described in Sect. 9.1, a class of optical planar circuits is the waveguide type, in which light energy is guided along straight or curved channels. This is, practically speaking, the most important class among various types of optical planar circuits in optical IC (integrated circuit) technology [10.1].

Three basic types of the waveguiding stripe structures are shown in Fig. 10.1. Structures in Fig. 10.1a, b, and c are called the ridge type, the embedded type, and the loaded type, respectively. The relations between the refractive indices are also shown in the figure. In a sense, this class of optical circuit should be classified as Category 5 in Table 1.1, because the width of the waveguiding structure is comparable (typically $5 - 20$ times) to the wavelength of light.

Therefore, the waveguiding behavior of such structures will be dealt with in this chapter only briefly, because the principal subjects of this book are microwave and optical planar circuits. We will mainly consider planar struc-

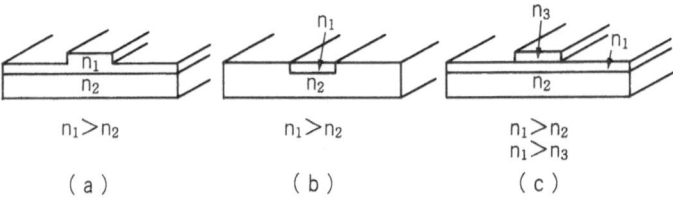

$n_1 > n_2$ $n_1 > n_2$ $n_1 > n_2$
$n_1 > n_3$

(a)   (b)   (c)

Fig. 10.1a – c. Three basic types of waveguiding stripe structures: (**a**) ridge type, (**b**) embedded type, (**c**) loaded type

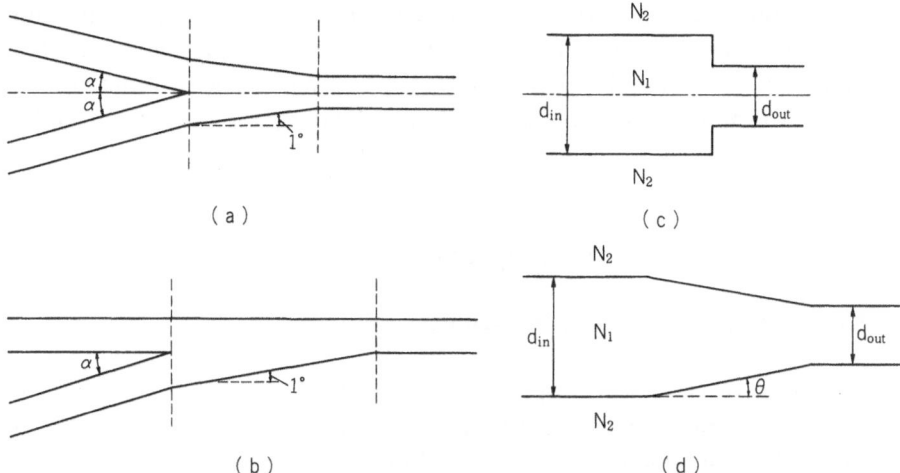

Fig. 10.2a–d. Examples of optical planar circuits having stripelike waveguiding structures: **(a)** symmetrical straight-guide Y junction, **(b)** non-symmetrical straight-guide Y junction, **(c)** abrupt transition, **(d)** tapered transition [10.2]

tures such as those shown in Fig. 10.2 [10.2], which consist basically of the stripelike waveguiding structures that, as the result of their combination, have a "planar" nature.

10.2 Model to be Considered

The model to be considered is essentially a two-dimensional one based on the concept of the effective refractive index defined in Sect. 9.2.2. The basic idea is illustrated in Fig. 10.3 for the ridge-type structure as an example. Different effective refractive indices N_1 and N_2 can be defined for Regions I and II of Fig. 10.3a by using relations derived in Sects. 9.2.2 (TE modes) and 9.2.3 (TM modes), as shown in Fig. 10.3b. Note that Fig. 10.3b does not show a "plate", but the height is infinite. Thus, the lightwave behavior in all the optical circuits shown in Fig. 10.2 can be dealt with as an entirely two-dimensional problem.

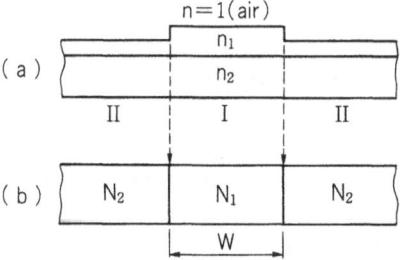

Fig. 10.3. (a) A ridge-type waveguiding structure, and **(b)** its "effective refractive-index model"

In the following, we first consider the basic properties of such a waveguiding structure very briefly using the geometrical optics approach (Sect. 10.3) and wave-optics approaches (Sect. 10.4). The analysis of the two-dimensional circuit performance is considered in Sect. 10.5 by using the beam-propagation method (BPM).

10.3 Geometrical Optics Approach

10.3.1 Limitation of Geometrical Optics

The concepts of geometrical optics and a ray, introduced in Sect. 9.3.1, are useful in analyzing the behavior of light in the structure shown in Fig. 10.3b, but under limited conditions as described in the following.

When the effective index $N(x, z)$ is a gradual function of x and z, as suggested in Sect. 9.3.2, the concepts of eikonal and ray are applicable provided that the spatial variation of the effective refractive index is gradual compared to the wavelength λ. Ordinary optical waveguiding structures may or may not satisfy such conditions.

When $N(x, z)$ is uniform in the stripe and outside regions, and a discrete boundary exists between these regions, as in Fig. 10.3b, an important parameter that determines the precision of the geometrical optics solution is the stripe width W (Fig. 10.3). When W is much larger than the wavelength (i.e., $W > 100\,\lambda$), the geometrical optics and hence the ray concept can be applied with practically satisfactory accuracy. When $W \approx 10\,\lambda$, as in many practical cases, the geometrical optics and ray concepts are still useful, but the correction due to of Goos-Haenchen shift [10.3, 4] (a phase shift that takes place at each total reflection of ray at the waveguide outside boundaries) is necessary to assure a practical accuracy. When the width W is so small as to permit only one propagating mode (single-mode waveguide), the ray concept is usually not usuable.

However, detailed description of the above conditions is outside the scope of this book; in this section, only a very brief description of the approximate analysis of the propagating condition follows.

10.3.2 Propagating Condition

Consider a waveguide consisting of a "uniform-core" (the waveguiding region) with width W and effective refractive index N_1, and "cladding" (the outside region) with effective refractive index N_2, where $N_2 < N_1$ (Fig. 10.4).

A ray is incident on the center of an end of the core of the waveguide with angle θ_0, as shown in the figure. It enters the waveguide, after refraction at the air-core boundary, with the angle θ given by Snell's law as

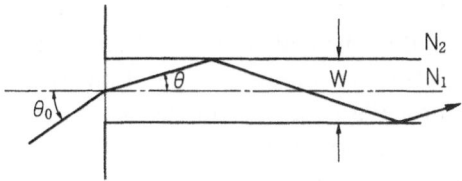

Fig. 10.4. A ray propagating in the core part of a uniform-core waveguiding structure

$$\sin \theta = \sin \theta_0 / N_1 . \tag{10.1}$$

On the other hand, for the ray to travel in the core subject to total reflections repeatedly at the core-cladding boundary, from Snell's law,

$$\sin \theta < \sqrt{N_1^2 - N_2^2} / N_1 . \tag{10.2}$$

Hence, a ray can travel in the waveguide only when the incident angle θ_0 satisfies [10.5]

$$\sin \theta_0 < \sqrt{N_1^2 - N_2^2} . \tag{10.3}$$

In the discussion of uniform-core waveguides, the relative refractive-index difference defined as

$$\Delta \equiv (N_1^2 - N_2^2)/2N_1^2 \simeq (N_1 - N_2)/N_1 \tag{10.4}$$

is often used. By using this parameter, (10.3) is rewritten as

$$\sin \theta_0 < N_1 \sqrt{2\Delta} . \tag{10.5}$$

Equation (10.5) is the basic formula giving the propagating condition of the waveguide. Note that both N_1 and Δ are functions of the mode number m (9.13, 16).

10.4 Wave-Optics Approaches

Wave-optics approaches for the analysis of stripelike optical waveguiding structures can be classified into two classes.

The first class includes three-dimensional wave analyses of the waveguide. In this category we have an approximate analysis technique proposed by *Marcatili* in 1969 [10.6]; various vectorial-wave, finite-element method (FEM) analyses over the waveguide cross section [10.7 – 10]; and mode-matching method analysis [10.1]. Recently the FEM analysis has been extended to waveguides using various anisotropic and gyrooptic materials. A large number of papers have appeared in this area; however, detailed description of such works is beyond the scope of this book, which is intended to cover principally the planar circuit problems.

The second class is basically a two-dimensional analysis; that is, the waveguide structure is represented by a two-dimensional, but essentially stripe-shaped distribution of the effective refractive index $N(x, z)$. Essentially, various mathematical approaches are possible to deal with the behavior of optical waves in such structures. However, most of the methods used in the microwave region are not practical in the optical frequencies because the width of the stripe is usually much wider (more than ten times greater) than the wavelength, and circuit length (for example, the length of a taper waveguide or a coupler) is usually as long as $100 - 1,000$ times the wavelength. As a result, the only practical mathematical approach reported so far is the beam-propagation method (often called BPM), developed by *Feit* and *Fleck* [10.12, 13]. In the following section, the principle and applications of the beam-propagation method to optical planar circuits having stripelike structures are described.

10.5 Beam-Propagation Method

The following description is essentially the two-dimensional version of the derivation shown in [10.12], which considers a three-dimensional beam propagation.

10.5.1 Principle

We start with the approximate two-dimensional scalar-wave equation (9.9):

$$\left(\frac{\partial^2}{\partial x^2} + \frac{\partial^2}{\partial z^2}\right) X + k_0^2 N^2(x, z) X = 0 , \tag{10.6}$$

where $N(x, z)$ denotes the two-dimensional effective refractive-index distribution, and X denotes the (x, z) dependence of the transverse magnetic field as

$$H_y(x, y, z) = Y(y)X(x, z) \tag{10.7}$$

for the TE modes which are assumed in (9.9). For TM modes, we may consider that $X(x, y)$ denotes the same for the transverse electric field as

$$E_y(x, y, z) = Y(y)X(x, z) . \tag{10.8}$$

Thus, these two cases are unified to give the same mathematical problem of solving (10.6).

In the present case we are only concerned with a wave propagating almost parallel to the z axis, because the optical planar circuit is assumed to have a stripelike structure. If we write, therefore,

$$X \doteq X_1 e^{-j\beta(z)z}, \tag{10.9}$$

we obtain from (10.6)

$$\left[\frac{\partial^2}{\partial x^2} - \beta^2(z) + k_0^2 N^2 \right] X = 0. \tag{10.10}$$

We further write

$$\partial^2/\partial x^2 = \nabla_\tau^2, \tag{10.11}$$

and forget for a while the fact that this is not a mathematical quantity but an operator. Then we may write "formally"

$$\beta(z) = [\nabla_\tau^2 + k_0^2 N^2(x, z)]^{1/2}. \tag{10.12}$$

Hence, the wave amplitude at $z + dz$ is given as

$$X(x, z + \Delta z) = \exp\{-j \Delta z [\nabla_\tau^2 + k_0^2 N^2(x, z)]^{1/2}\} X(x, z). \tag{10.13}$$

We may, however, rewrite the square root in the right-hand side as

$$(\nabla_\tau^2 + k_0^2 N^2)^{1/2} = \frac{\nabla_\tau^2}{(\nabla_\tau^2 + k_0^2 N^2)^{1/2} + k_0 N} + k_0 N. \tag{10.14}$$

This relation can easily be proved by a simple algebraic manipulation.

The essential point in the BPM lies in the following approximation, in which $N(x, z)$ in the denominator of the first term of (10.14) is approximated by a constant N_0, typically the value in the substrate of the circuit, so that

$$(\nabla_\tau^2 + k_0^2 N^2)^{1/2} \doteq \frac{\nabla_\tau^2}{(\nabla_\tau^2 + k_0^2 N_0^2)^{1/2} + k_0 N_0} + k_0 N_0 + k_0 N_0 \left(\frac{N}{N_0} - 1 \right). \tag{10.15}$$

This approximation gives satisfactory accuracy when an inequality $(N - N_0)/N_0 \ll 1$ holds; the proof is shown in Ref. [10.13], Appendix.

We next write

$$X(x, z) = W(x, z) e^{-jk_0 N_0 z}. \tag{10.16}$$

Putting the above equation into (10.13) and using the approximation used in (10.15), we have

$$W(x, z + \Delta z) = \exp\left[-j \Delta z \left(\frac{\nabla_\tau^2}{(\nabla_\tau^2 + k_0^2 N_0^2)^{1/2} + k_0 N_0} + \chi(x, z) \right) \right]$$
$$\times W(x, z) + O^3(\Delta z), \tag{10.17}$$

where

$$\chi(x, z) = k_0 N_0 \left(\frac{N}{N_0} - 1 \right) , \tag{10.18}$$

and the final term gives the remaining computational error [10.12].

Equation (10.17) can further be rewritten as

$$W(x, z + \Delta z) = \exp \left(\frac{-j\Delta z}{2} \frac{\nabla_\tau^2}{(\nabla_\tau^2 + k_0^2 N_0^2)^{1/2} + k_0 N_0} \right) \exp(-j\Delta z \chi)$$

$$\times \exp \left(\frac{-j\Delta z}{2} \frac{\nabla_\tau^2}{(\nabla_\tau^2 + k_0^2 N_0^2)^{1/2} + k_0 N_0} \right) W(x, z) + O^3(\Delta z) . \tag{10.19}$$

We consider here the physical model of (10.19). The first factor in (10.19),

$$\exp \left(-\frac{j\Delta z}{2} \frac{\nabla_\tau^2}{(\nabla_\tau^2 + k_0^2 N_0^2)^{1/2} + k_0 N_0} \right) ,$$

is equivalent to solving a wave equation in a homogeneous medium

$$\left(\frac{\partial^2}{\partial x^2} + \frac{\partial^2}{\partial z^2} + k_0^2 N_0^2 \right) X = 0 , \tag{10.20}$$

for a distance ($\Delta z/2$), using $X(x, z)$ [$= W(x, z)$] as the initial condition. This equivalence can be proved easily by substituting $\chi = 0$ into (10.17). Therefore, the three factors (exponential operators) in (10.19) can be represented by a homogeneous medium (length $\Delta z/2$), a lens giving the phase shift $-j\Delta z\chi(x, z)$, and again a homogeneous medium (length $\Delta z/2$). This means that advancing the solution for $W(x, z)$ by repeated application of (10.19) is equivalent to propagating the light beam through a periodic array of thin optical lenses as shown in Fig. 10.5 [10.12]. The first lens is located at $z = \Delta z/2$. The second, third, and further lenses are separated from one another by the distance Δz. Each lens gives an x-dependent phase shift given by $\Delta z\chi(x, z)$ to the beam, whereas the beam propagation between lenses is governed by (10.20). Thus, the propagation of the beam can be traced in a step-by-step manner.

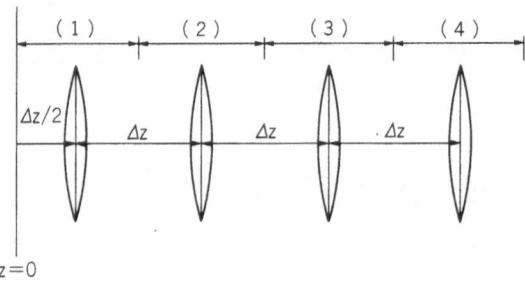

Fig. 10.5. An array of lenses equivalent to the beam-shape transformation expressed by (10.19). One section consists of a uniform medium with a length ($\Delta z/2$), a thin lens, and a uniform medium with a length ($\Delta z/2$) [10.12]

10.5.2 Numerical Calculations

An alternative representation of (10.19), which is more suitable for numerical calculations, can be obtained by expressing $W(x, z)$ as a Fourier series having a finite number of terms:

$$W(x, z) = \sum_{n=-N/2}^{N/2} W_n(z) \exp(jk_{xn}x), \tag{10.21}$$

where k_{xn}'s denote discrete transverse wavenumbers defined as

$$k_{xn} = \frac{2\pi}{L} n, \tag{10.22}$$

whereas L is the width of the computational area.

We consider one section in Fig. 10.5 which consists of a homogeneous medium having an effective index of N_0 and length $(\Delta z/2)$, a thin lens, and again a homogeneous medium of length $(\Delta z/2)$. The nth Fourier component W_n at $z+(\Delta z/2)$ can be obtained, by substituting $\chi(x, z) = 0$ into (10.17) and using $\nabla_\tau^2 = -k_{xn}^2$, as

$$W_n\left(z+\frac{\Delta z}{2}\right) = W_n(z) \exp\left(\frac{j\Delta z}{2} \frac{k_{xn}^2}{(-k_{xn}^2+k_0^2N_0^2)^{1/2}+k_0N_0}\right). \tag{10.23}$$

We may then reconstruct the real-space function W in the real space just behind the thin lens, that is, $W(x, z+(\Delta z/2)-0)$, by using the Fourier components $W_n(z+(\Delta z/2))$. The actual computation can be performed by using the widely available FFT (Fast Fourier Transform) algorithm [10.14]. Then, multiplying it by the "lens" term $\exp[-j\Delta z\chi(x, z)]$ in (10.19), we obtain the function just in front of the thin lens: $W(x, z+(\Delta z/2)+0)$. The beam propagation in the following homogeneous space of length $(\Delta z/2)$ can be calculated again by (10.23) and two FFT processes.

Thus, the beam propagation over a single section is calculated. Of course, the final FFT can be omitted if the beam shape (power distribution) at this point is not needed, and the calculation is to be continued to the following section.

10.5.3 Examples of Calculation Results

Some results of the BPM calculations that have appeared in the literature [10.2] are shown in the following.

Figure 10.6 shows the propagation of a light beam in a tapered waveguide (from bottom to top), for four input modes. The height of distribution pat-

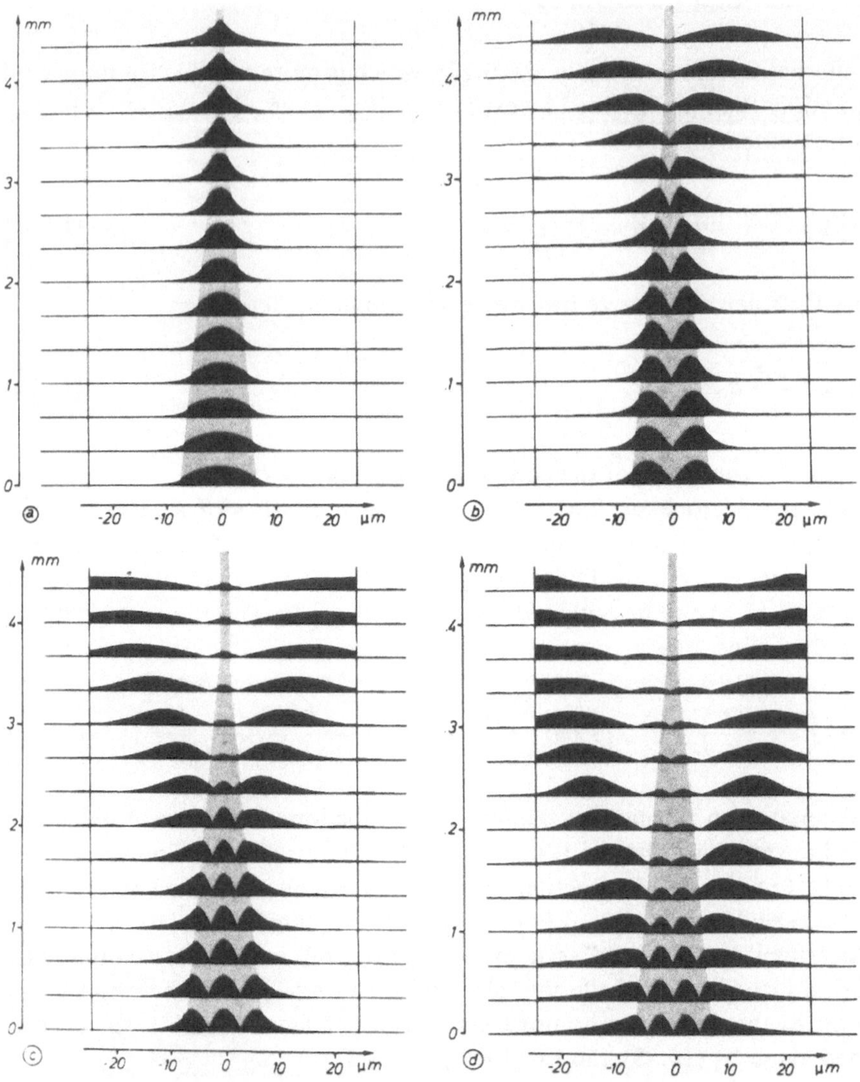

Fig. 10.6a – d. Propagation of a light beam in a tapered waveguide. The beam travels from the bottom to the top of the figure. Note that the abscissa is expanded about 5 times relative to the ordinate: **(a)** zero-order mode, **(b)** first-oder mode **(c)** second-order mode, **(d)** third-order mode [10.2]

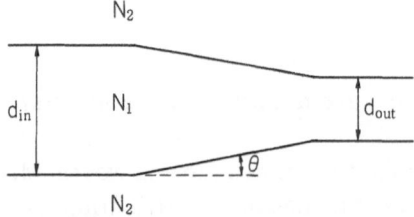

Fig. 10.7. Configuration and geometry of the tapered section assumed in the analyses whose results are shown in Fig. 10.6 [10.2]

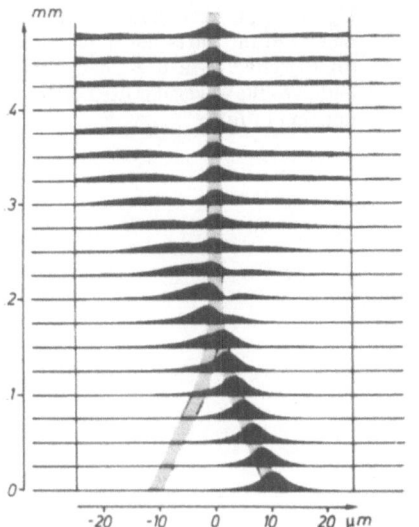

Fig. 10.8. Propagation of a light beam in a waveguide Y junction. The beam travels from the right bottom to the top of the figure. Note that the abscissa is expanded about 5 times relative to the ordinate [10.2]

terns gives the light amplitude. The effective refractive indices of the waveguide and substrate, N_1 and N_2, are assumed to be

$$N_2 = 1.00, \quad N_1^2 - N_2^2 = 0.02,$$

whereas the waveguide width at the input and output ends, d_{in} and d_{out} (Fig. 10.7), are given in terms of their normalized values as

$$V_{in} = k_0 d_{in}(N_1 - N_2)^{1/2} = 10.0,$$

$$V_{out} = k_0 d_{out}(N_1^2 - N_2^2)^2 = 1.25,$$

where k_0 is the wavenumber in vacuum. The taper angle θ (Fig. 10.7) is 1 degree, and the wavelength of the light beam is assumed to be 1.30 μm.

It is found how the light power is transferred to the radiation mode, especially when the mode number is high. Quantitative analyses of the radiation loss and mode conversion in such tapers are also performed in [10.2] on the basis of the BPM analysis results.

Figure 10.8 shows the light amplitude distribution in a Y junction (Fig. 10.9a), where $\alpha_1 = 2$ degrees, $\alpha_2 = 4$ degrees, the wavelength $\lambda = 1.3$ μm, and $A = 8\lambda$. As shown in the figure, the waveguide shape is given as

$$Z = x \tan \alpha_1 + (A - L \tan \alpha_1) \sin \frac{\pi x}{2L}. \tag{10.24}$$

The waveguide width is 6 μm in the entire circuit pattern; however, the effective refractive index is not uniform but has a smooth distribution as shown in

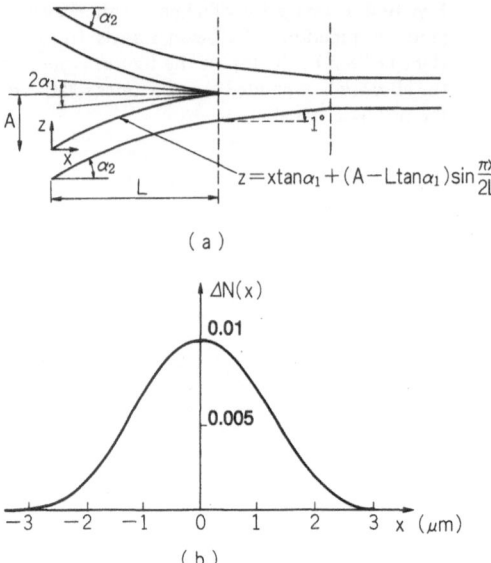

Fig. 10.9. (a) Configuration and geometry of the waveguide Y junction. **(b)** Effective refractive-index distribution of the waveguide used in the Y junction of **(a)**

Fig. 10.9b, where N_1 (value at the waveguide center) and N_2 are again given as $N_2 = 1.00$ and $N_1^2 - N_2^2 = 0.02$.

Figure 10.8 suggests that the input light power is apt to travel straight; an appreciable part of the power seems to be converted to radiation modes. In [10.2], the quantitative analyses of the radiation loss are performed for various circuit configurations upon the basis of the BPM analysis results such as shown in Fig. 10.8.

10.6 Summary

The analysis of optical planar circuits having stripelike waveguiding structures has been described. As readers are probably aware, the analysis technique in this area is not yet matured; the beam-propagation method (BPM) is the only efficient approach suited to this sort of circuit. This immaturity, on the other hand, offers us a challenging field for research in the future.

Appendix

A2.1 Derivation of (2.5)

We may derive (2.5), without loss of generality, from the calculated complex transmission constant of a parallel-plate transmission line as shown in Fig. A2.1. Hereafter the width and spacing of the line are denoted by a and d, respectively.

A2.1.1 Transmission Power

The power transmitted by the line is expressed, in terms of the complex Poynting vector in the y direction defined as $S_y = E_z H_x^*/2$, as

$$P = \int_0^d \int_0^a \mathrm{Re}\{S_y\}\, dx\, dz = ad\,|E_z|^2/2Z_0 , \tag{A2.1}$$

where $Z_0 = \sqrt{\mu/\varepsilon}$ denotes the wave impedance of the space.

A2.1.2 Conductor Loss

Next, we compute the conductor loss assuming that small conductor loss (small enough not to disturb the field configuration) exists on the surface of conductors. When such loss exists, the Poynting vector has a small component S_n normal to the conductors. Then, the power attenuation rate is given as

$$-\frac{dP}{dz} = 2a\,\mathrm{Re}\,\{S_n\} . \tag{A2.2}$$

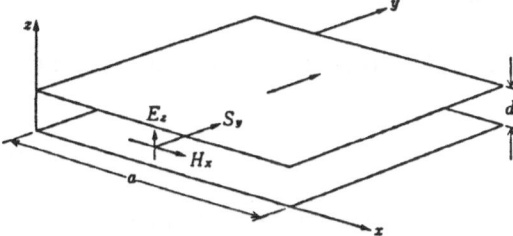

Fig. A2.1. Symbols used in the computation of the attenuation in a planar circuit

On the other hand, the characteristic impedance of a metallic material is given, when $\sigma \gg \omega\varepsilon$, as

$$Z_m = \sqrt{\frac{j\omega\mu}{\sigma + j\omega\varepsilon}} \doteq \sqrt{\frac{\omega\mu}{2\sigma}} (1 + j) . \tag{A2.3}$$

Therefore,

$$2a\,\mathrm{Re}\{S_n\} = a\,\mathrm{Re}\{Z_m H_x H_x^*\} = a\,|H_x|^2 \sqrt{\omega\mu/2\sigma} = a\omega\mu r\,|H_x|^2/2 , \tag{A2.4}$$

where r denotes the skin depth of the conductor, which is given as

$$r = \sqrt{2/\omega\mu\sigma} . \tag{A2.4'}$$

The attenuation constant due to the conductor loss, α_c, is defined as

$$-\frac{dP}{dz} = 2\,\alpha_c P . \tag{A2.5}$$

Hence, from (A2.1 – 5), we have

$$\alpha_c = \frac{-(dP/dz)}{2P} = \frac{a\omega\mu r\,|H_x|^2/2}{2ad\,|E_z|^2/2Z_0} = \omega\sqrt{\varepsilon\mu}\,\frac{r}{2d} . \tag{A2.6}$$

A2.1.3 Dielectric Loss

The power attenuation rate due to the dielectric loss is expressed as

$$-\frac{dP}{dz} = \int_0^d \int_0^a \frac{1}{2}(\omega\varepsilon\tan\delta)\,|E_z|^2\,dx\,dz , \tag{A2.7}$$

where δ denotes the loss angle of the spacing material $[= \tan^{-1}(\sigma/\omega\varepsilon)]$. From (A2.1 – 7), the attenuation constant due to the dielectric loss, α_d, is given as

$$\alpha_d = \omega\varepsilon Z_0\tan\delta/2 = \omega\sqrt{\varepsilon\mu}\,\tan\delta/2 . \tag{A2.8}$$

A2.1.4 Complex Propagation Constant

Equations (A2.6, 8) tell us that the complex propagation constant of the line is given as

$$\beta = \omega\sqrt{\varepsilon\mu}\,[1 - j\,(\tan\delta + r/d)/2] . \tag{A2.9}$$

Using this modified β is equivalent to considering formally that the wave-number k is replaced by (2.5a) in the text.

A2.2 Some Characteristics of Eigenvalues and Eigenfunctions

The eigenvalue problem of a planar circuit is identical to that of TE-mode waves in a hollow metallic waveguide [2.2, 6][1]. This identity can readily be understood by replacing V by H_z, and k^2 by β^2 in (2.9). Thus the present eigenvalue problem has a rather long history as a mathematical problem in microwave engineering.

A2.2.1 Formula Giving k^2

Generally, the eigenvalue problem is formulated as

$$(\nabla_T^2 + k^2)\phi_n = 0, \qquad k^2 = \omega^2 \varepsilon \mu, \tag{A2.10a}$$

$$\partial\phi_n/\partial n = 0 \quad (\text{on } C), \tag{A2.10b}$$

where ϕ_n denotes the nth eigenfunction. We derive first the variational expression of the eigenvalues (2.11).

Multiplying (A2.10a) by ϕ_n^* and integrating over Region D, we obtain, using the two-dimensional Gauss's theorem,

$$\iint_D (\phi_n^* \nabla^2 \phi_n + k^2 |\phi_n|^2) \, dS$$

$$= \oint_C \phi_n^* \frac{\partial\phi_n}{\partial n} \, ds - \iint_D |\nabla\phi_n|^2 \, dS + k^2 \iint_D |\phi_n|^2 \, dS = 0. \tag{A2.11}$$

Combining this equation with the boundary condition (A2.10b), we obtain

$$k^2 = \frac{\iint_D |\nabla\phi_n|^2 \, ds}{\iint_D |\phi_n|^2 \, ds}. \tag{A2.12}$$

Here k^2 is always positive (including zero).

A2.2.2 Proof that (2.11) Gives a Stationary Expression

Next we prove that (2.11) in the text or (A2.12) is a stationary expression of k^2.

We assume that when ϕ_n varies slightly to $\phi_n + \delta\phi_n$, k^2 becomes $k^2 + \delta k^2$. Considering only the first-order terms of both $\delta\phi_n$ and its derivative, we have

$$\delta k^2 \iint_D \phi_n^2 \, dS = 2 \iint_D (\nabla\phi_n \nabla\delta\phi_n - k^2 \delta\phi_n \phi_n) \, dS. \tag{A2.13}$$

Using further the two-dimensional Gauss's theorem, we obtain

---

[1] References cited are mentioned in the corresponding chapters

$$\delta k^2 \iint_D \phi_n^2 dS = 2 \oint_C \delta\phi_n \frac{\partial\phi_n}{\partial n} ds - 2\iint_D \delta\phi_n(\nabla^2\phi_n + k^2\phi_n) dS. \qquad (A2.14)$$

Substituting (A2.10a, b) into (A2.14), we see that the first-order variation δk^2 is zero; i.e., (A2.12) is a stationary expression of k^2.

A2.2.2 Orthogonality of Eigenfunctions

We first multiply (A2.10a) by ϕ_m. Next, we make a similar product but exchange n and m. By integrating the difference of these two products over Region D, we have

$$\iint_D (\phi_m\nabla^2\phi_n - \phi_n\nabla^2\phi_m) dS = (k_m^2 - k_n^2) \iint_D \phi_n\phi_m dS. \qquad (A2.15)$$

By using Green's theorem and the boundary condition, we can show that the left-hand side is zero. Thus we see that for two modes having different eigenvalues (i.e., $k_m \neq k_n$), the orthogonality holds [(2.12) in the text].

A3.1 Weber's Solution Using Cylindrical Waves

We consider a point P in a two-dimensional region D inside a contour C, as shown in Fig. A3.1. It is shown in [3.2] that the rf potential $V(P)$ satisfying the Helmholtz equation can be expressed in terms of V and its normal derivative $\partial V/\partial n$ along the contour C. In [3.2], Neumann function $Y_0(kr)$ is used as the Green's function. However, function $Y_0(kr)\exp(j\omega t)$ expresses a sort of standing cylindrical wave [3.3]. In the present case, instead of $Y_0(kr)\exp(j\omega t)$, the Hankel function of the second kind, $H_0^{(2)}(kr)\exp(j\omega t)$, which expresses a diverging cylindrical wave, seems more appropriate. In the following we rewrite Weber's solution by using the Hankel function. A more general form of the Green's function, including both $Y_0(kr)$ and $H_0^{(2)}(kr)$ as

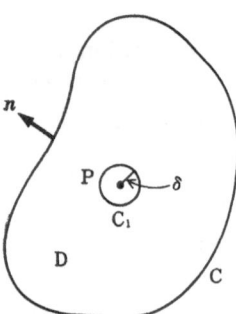

Fig. A3.1. Symbols used in the Weber's solution

ties, has been proposed by *Ayasli* [3.19] and discussed by *Kitazawa* [3.20]. This problem is discussed in Appendix A3.3.

We now consider two functions v and w which satisfy

$$(\nabla^2 + k^2)\, V = 0 \tag{A3.1}$$

in D, where V stands for v and w. From Green's theorem, we have

$$\oint_C \left(v\, \frac{\partial w}{\partial n} - w\, \frac{\partial v}{\partial n} \right) ds = \iint_D (v\nabla^2 w - w\nabla^2 v)\, dS = 0. \tag{A3.2}$$

Here we assume that $v = V$, and that

$$w = H_0^{(2)}(kr), \tag{A3.3}$$

where r denotes the distance between Point P and a point on the contour C.

When Point P is located outside the contour C, from (A3.2),

$$\oint_C \left[V\, \frac{\partial H_0^{(2)}(kr)}{\partial n} - H_0^{(2)}(kr)\, \frac{\partial V}{\partial n} \right] ds = 0 \tag{A3.4}$$

holds. When Point P is inside the contour C, P becomes a singular point for w, and the area integral in (A3.2) does not vanish. In this case, we apply the Green's theorem to the region excluding a small circular area (radius δ) around P surrounded by contour C_1, to obtain

$$\oint_C \left[V\, \frac{\partial H_0^{(2)}(kr)}{\partial n} - H_0^{(2)}(kr)\, \frac{\partial V}{\partial n} \right] ds = \oint_{C_1} \left[V\, \frac{\partial H_0^{(2)}(kr)}{\partial r} - H_0^{(2)}(kr)\, \frac{\partial V}{\partial r} \right] ds. \tag{A3.5}$$

Here we have used the relation $\partial n = -\partial r$ on C_1.

We use now the following approximations

$$H_0^{(2)}(kr) \doteq -\frac{2\mathrm{j}}{\pi} \log \frac{kr}{2}, \qquad \frac{\partial H_0^{(2)}(kr)}{\partial r} \doteq -\frac{2\mathrm{j}}{\pi r}, \tag{A3.6}$$

which hold for $kr \ll 1$, and assume for a while that V and $\partial V/\partial n$ are constant on C_1. Then we may rewrite the right-hand side of (A3.5) as

$$\int_{C_1} \left(-\frac{2\mathrm{j}}{\pi r}\, V + \frac{2\mathrm{j}}{\pi} \log \frac{kr}{2}\, \frac{\partial V}{\partial n} \right) ds = -4\mathrm{j}\, V + 4\mathrm{j}\,\delta \log \frac{k\delta}{2}\, \frac{\partial V}{\partial n}. \tag{A3.7}$$

If we assume here further that $\delta \to 0$, the second term of the right-hand side vanishes. (Actually, $\partial V/\partial n$ is not constant but takes positive and negative values around Point P. Therefore, the second term of the right-hand side tends to zero more strongly.)

Thus, we obtain, for a point inside the contour C,

$$4\mathrm{j}\, V(P) = \oint_C \left[H_0^{(2)}(kr)\frac{\partial V}{\partial n} - V\frac{\partial H_0^{(2)}(kr)}{\partial n} \right] ds. \qquad (A3.8)$$

This is the starting equation giving the rf potential on Point P in the region D in terms of V and $\partial V/\partial n$ upon the contour C.

A3.2 Derivation of (3.1)

The rf potential at any point inside C can now be given as (A3.8). However, in the present analysis we need the rf potential "upon" C, which is derived in the following.

We set Point P' just on the contour C and Point P just inside C, as shown in Fig. A3.2, and assume that $\delta \ll \alpha \ll k^{-1}$. Using asymptotic expressions of Hankel functions for small argument

$$H_0^{(2)}(kr) \doteqdot -\frac{2\mathrm{j}}{\pi}\log\frac{k\sqrt{s^2+\delta^2}}{2}, \qquad (A3.9a)$$

$$\frac{\partial H_0^{(2)}(kr)}{\partial n} \doteqdot -\frac{2\mathrm{j}}{\pi}\frac{\delta}{s^2+\delta^2}, \qquad (A3.9b)$$

we obtain, from (A3.8),

$$4\mathrm{j}\, V(P) = \int_{-\alpha}^{\alpha} \left(-\frac{2\mathrm{j}}{\pi}\log\frac{k\sqrt{s^2+\delta^2}}{2}\frac{\partial V}{\partial n} + \frac{2\mathrm{j}}{\pi}\frac{\delta}{s^2+\delta^2}V \right) ds$$

$$+ \int_\Gamma \left[H_0^{(2)}(kr)\frac{\partial V}{\partial n} - V\frac{\partial H_0^{(2)}(kr)}{\partial n} \right] ds. \qquad (A3.10)$$

In the above equation, Γ denotes the contour excluding the infinitesimal region between $-\alpha$ and $+\alpha$ (Fig. A3.2).

If we assume here that V and $\partial V/\partial n$ do not change appreciably in the region $-\alpha \sim +\alpha$, the integrals in (A3.10) will become

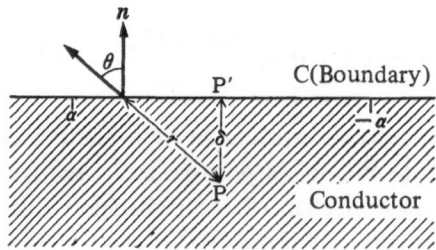

Fig. A3.2. Symbols used in the derivation of the integral equation

$$-\frac{2j}{\pi}\int_{-\alpha}^{\alpha}\log\frac{k\sqrt{s^2+\delta^2}}{2}\frac{\partial V}{\partial n}\,ds = -\frac{4j}{\pi}\frac{\partial V}{\partial n}$$

$$\times\left[\alpha\left(\log\frac{k\sqrt{\alpha^2+\delta^2}}{2}-1\right)-\frac{k\delta}{2}\left(\mathrm{arccosec}\frac{\sqrt{\alpha^2+\delta^2}}{\delta}-\frac{\pi}{2}\right)\right],$$

$$(A3.11)$$

$$\frac{2j}{\pi}\int_{-\alpha}^{\alpha}\frac{s}{s^2+\delta^2}V\,ds = \frac{4j}{\pi}V\tan^{-1}\frac{\alpha}{\delta}. \tag{A3.12}$$

We assume next that P approaches P' so that $\delta\to 0$. Then, (A3.10–12) lead to

$$4j\,V(P') = -\frac{4j}{\pi}\frac{\partial V}{\partial n}\left[\alpha\left(\log\frac{k\alpha}{2}-1\right)\right]+2j\,V(P')$$

$$+\int_{\Gamma}\left[H_0^{(2)}(kr)\frac{\partial V}{\partial n}-V\frac{\partial H_0^{(2)}(kr)}{\partial n}\right]ds. \tag{A3.13}$$

By assuming further that $\alpha\to 0$, the first term of the right-hand side vanishes, so that

$$2j\,V(P') = \oint_C\left[H_0^{(2)}(kr)\frac{\partial V}{\partial n}-V\frac{\partial H_0^{(2)}(kr)}{\partial n}\right]ds. \tag{A3.14}$$

Combining the above equation with the relations

$$\frac{\partial V}{\partial n} = -j\,\omega\mu\,di_n, \tag{A3.15}$$

$$\frac{\partial H_0^{(2)}(kr)}{\partial n} = -k\cos\theta\,H_1^{(2)}(kr), \tag{A3.16}$$

we obtain (3.1).

A3.3 Choice of the Green's Function Used in Contour-Integral Analysis

In the analysis described in Sect. 3.2, the zero-order Hankel function of the second kind $H_0^{(2)}(kr)$ $[=J_0(kr)-j\,Y_0(kr)]$ and its derivative $-H_1^{(2)}(kr)$ appear in the starting contour-integral equation (3.1). This is because, as seen in its derivation in Appendix A3.1 (A3.1–3), $H_0^{(2)}$ is employed as the Green's function of the two-dimensional free space.

The function $H_0^{(2)}(kr)$ expresses a diverging cylindrical wave, whereas the same function of the first kind $H_0^{(1)}(kr)$ $[= J_0(kr) + j Y_0(kr)]$ expresses a converging one [3.3]. Hence, the choice of $H_0^{(2)}$ seems to be natural. However, *Ayasli* considered that mathematically any linear combination of $H_0^{(1)}$ and $H_0^{(2)}$ should give the same result [3.19]. He proposed to use a more general expression

$$G = C_0 J_0(kr) - \tfrac{1}{4} Y_0(kr), \qquad (A3.17)$$

in which C_0 is equal to $-j/4$ in the analysis in Chap. 3, whereas it could (mathematically) be chosen more arbitrarily. On the other hand, Ayasli also asserted that when the analysis is finally performed in a matrix form, as is the case here, an appropriate choice of C_0 would improve the computational accuracy.

Figure A3.3 shows an example of his results. The ordinate shows the lowest resonant frequency f_{01} of a circular resonator, whereas the abscissa gives the absolute value of C_0 (real values are assumed). This figure apparently suggests the superiority of large (real, but either positive or negative) values of C_0.

Kitazawa investigated this problem further with special emphasis on the accuracy in a wider frequency range, i.e., above the lowest resonant frequency f_{01} [3.20]. The results indicate that the relation between C_0 and the accuracy is not so simple; the accuracy strongly depends upon the shape of the circuit to be analyzed. Besides, a large C_0 (either real or imaginary) often results in generation of spurious (physically nonexisting) solutions at frequencies above f_{01}. On the whole, $C_0 = -j/4$ is probably not a bad choice.

In *Kitazawa*'s analysis described in the following, a rectangular circuit as shown in Fig. A3.4 is assumed. The spacing material is assumed to have a specific permittivity of $\varepsilon_s = 2.53$. The external lines are assumed to have a

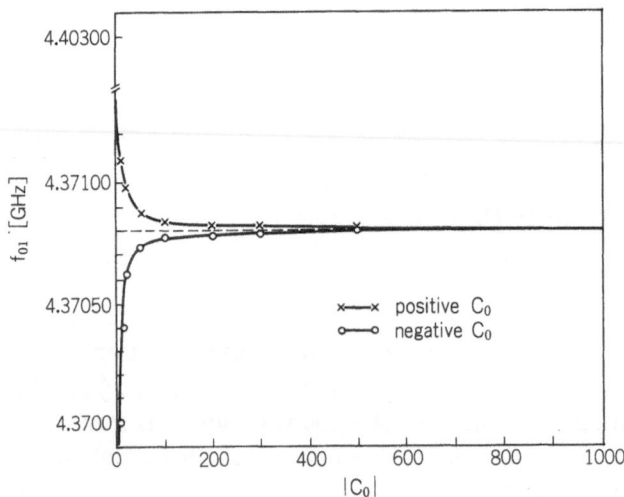

Fig. A3.3. The lowest resonant frequency of a circular resonator as functions of C_0 used in (A3.17) [3.19]

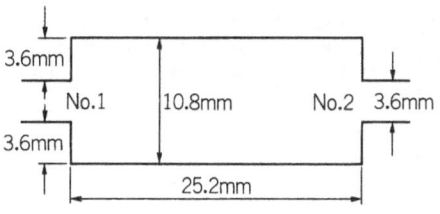

Fig. A3.4. A rectangular circuit used in Kitazawa's analysis. All dimensions are in millimeters. The circuit spacing is assumed to be 1.52 mm [3.20]

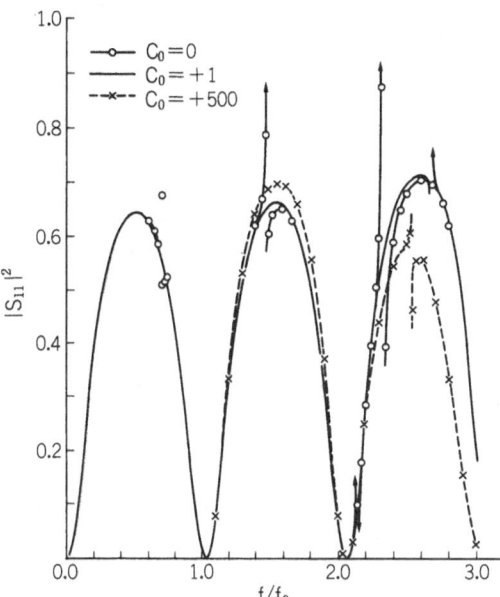

Fig. A3.5. Effect of the parameter C_0 on the computed frequency characteristics, for positive real values of C_0 [3.20]

characteristic impedance of 50. First, the values of matrix element S_{11} computed by using various C_0 are shown as functions of frequency.

A3.3.1 Cases when C_0 is Positive Real

The result is shown in Fig. A3.5. When C_0 is large, spurious solutions appear, in particular at frequencies above $2f_{01}$.

A3.3.2 Cases when C_0 is Negative Real

See Fig. A3.6. The spurious solutions appear at frequencies around or even below f_{01}.

A3.3.3 Cases when C_0 is Imaginary

The result is shown in Fig. A3.7, together with the computed $|S_{11}|^2 + |S_{21}|^2$ which should be a unity because of energy conservation. The results for

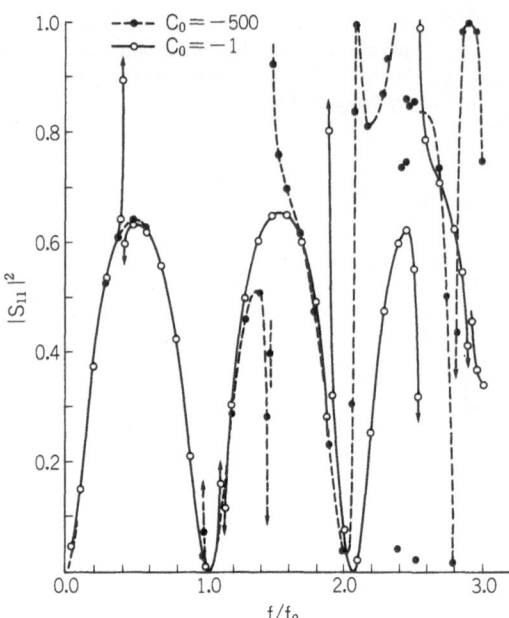

Fig. A3.6. Effect of the parameter C_0 on the computed frequency characteristics, for negative real values of C_0 [3.20]

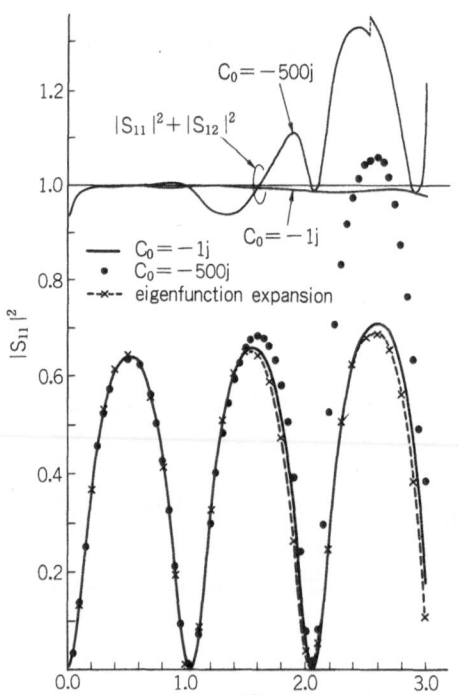

Fig. A3.7. Effect of the parameter C_0 on the computed frequency characteristics, for imaginary values of C_0 [3.20]

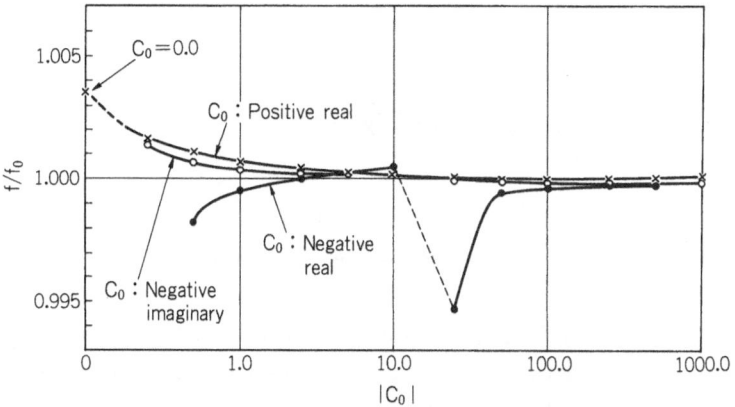

Fig. A3.8. Resonant frequency of a rectangular circuit (Fig. A3.4) as functions of $|C_0|$ for the lowest resonance [3.20]

$C_0 = -j/4$ and $C_0 = -j$ were practically identical. Small crosses on a broken curve show the result of eigenfunction-expansion analysis which is free from the spurious problem. It is found that when C_0 is imaginary, the spurious solution does not appear, but the accuracy is degraded at $f > 2f_0$ when $|C_0|$ is very large.

Finally, the accuracies of f_{01} and f_{02} computed from the first and second minima of $\det U$ (3.7) were investigated; the result for f_{01} is shown in Figs. A3.8 for the above three cases. The positive imaginary C_0 and negative imaginary C_0 give quite similar results. It has been found that a large $|C_0|$ gives better accuracy for f_{01}, but not always for f_{02}.

In conclusion, it has been found that the relation between C_0 and the computational accuracy is rather complex, and that the choice of $C_0 = -j/4$ is probably good, not only because it agrees best with the physical picture, but also because it gives relatively good accuracy in all the cases tested.

A4.1 Proof of (4.1) for a Multiply Connected Circuit Pattern

We now consider a planar circuit pattern having a hole, as shown in Fig. A4.1a. We connect the outer and inner boundaries, C_1 and C_2, with contours C_3 and C_4 which are infinitesimally separated, and give a direction of integration (Fig. A4.1b), to define θ. For the entire contour $C_1 + C_2 + C_3 + C_4$, obviously (4.1) in the text holds. However, for the corresponding points upon C_3 and C_4 (Fig. A4.1c),

$$[V(s_0)]_{C_3} = [V(s_0)]_{C_4}, \tag{A4.1}$$

$$[i_n(s_0)]_{C_3} = -[i_n(s_0)]_{C_4}, \tag{A4.2}$$

must be satisfied.

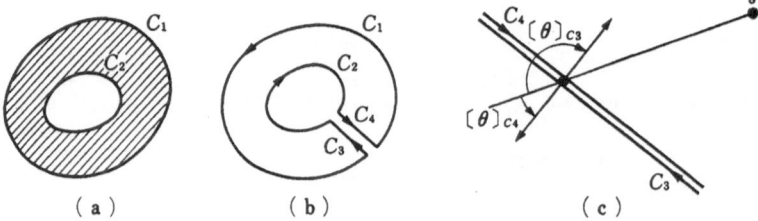

Fig. A4.1a–c. Proof of (4.1) for a multiply connected planar circuit pattern

The voltage $V(s)$ in the left-hand side of (4.1) needs to be known only upon C_1 and C_2. If we consider cases in which s is located somewhere upon C_1 or C_2, and s_0 is located somewhere upon C_3 and C_4, we obtain for the corresponding points upon C_3 and C_4

$$[\theta]_{C_3} = [\theta]_{C_4} + \pi . \tag{A4.3}$$

Hence,

$$[\cos \theta]_{C_3} = -[\cos \theta]_{C_4} . \tag{A4.4}$$

Putting (A4.1, 2, 4) into the right-hand side of (4.1), we find that the integrals along C_3 and C_4 cancel each other. Hence, we may apply (4.1) to a pattern like that in Fig. A4.1a, provided that the direction of integration along C_1 and C_2 is carefully defined.

A5.1 Computational Technique for Combining n Subports to a Single External Port

For simplicity, a case in which *two* subports (instead of n subports) are combined is considered [5.1, 2].

In Sect. 5.2.3 of the text, it was assumed that *Ports 1p and 2p are connected directly in the interface network to Ports k and l, respectively.* Now we consider that *Port 1p (for example) is connected in the interface to Ports k and k' in parallel.* This situation is depicted as a *local interface network* in Fig. A5.1. Here the externally connected terminal of the interface network is as-

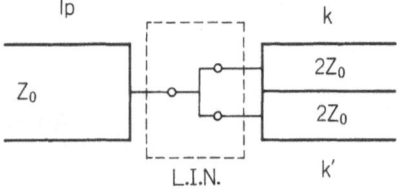

Fig. A5.1. Local interface network used at coupling ports

sumed to be connected to two sampling points on the circuit periphery; in other words, the external circuit has a width corresponding to two divisions along the periphery.

We assume that the voltages of the three terminals ($1p$, k, and k') are equal. Since the total sum of currents flowing into them is zero, the scattering matrix of the interface network depicted in Fig. A5.1 is given, by inspection, as

$$[S_N'] = \begin{pmatrix} 0 & 1/\sqrt{2} & 1/\sqrt{2} \\ 1/\sqrt{2} & -1/2 & 1/2 \\ 1/\sqrt{2} & 1/2 & -1/2 \end{pmatrix}. \tag{A5.1}$$

By inserting the above scattering matrix in appropriate locations in (5.4) of the text, the scattering matrix of the whole interface network can be obtained.

A8.1 Derivation of (8.5, 8)

A8.1.1 Derivation of (8.5)

From (8.1, 4a – d) in the text, we obtain

$$(E_z)_y = -j\omega\mu H_x - \omega\kappa H_y, \tag{A8.1}$$

$$-(E_z)_x = -j\omega\mu H_y + \omega\kappa H_x, \tag{A8.2}$$

$$\mu[(H_x)_x + (H_y)_y] - j\kappa[(H_y)_x - (H_x)_y] = 0, \tag{A8.3}$$

$$(H_y)_x - (H_x)_y = j\omega\varepsilon E_z, \tag{A8.4}$$

where ( )$_x$ and ( )$_y$ denote partial derivatives with respect to x and y, respectively. From (A8.1, 2), we obtain

$$(E_z)_{xx} + (E_z)_{yy} = j\omega\mu[(H_y)_x - (H_x)_y] - \omega\kappa[(H_x)_x - (H_y)_y]. \tag{A8.5}$$

Substituting (A8.3, 4) into (A8.5), we have

$$(E_z)_{xx} + (E_z)_{yy} = -\omega^2\varepsilon\mu_{\text{eff}}E_z, \quad \text{where} \tag{A8.6}$$

$$\mu_{\text{eff}} = (\mu^2 - \kappa^2)/\mu. \tag{A8.7}$$

A8.1.2 Derivation of (8.8)

Equations (A8.1, 2) yield

$$H_x = \frac{1}{\omega\mu_{\text{eff}}}\left[\frac{\kappa}{\mu}(E_z)_x - j(E_z)_y\right], \tag{A8.8}$$

$$H_y = \frac{1}{\omega\mu_{\text{eff}}} \left[\frac{\kappa}{\mu} (E_z)_y - \text{j}(E_z)_x \right]. \tag{A8.9}$$

Equation (8.8) can be derived from (A8.8, 9) in a manner similar to the one shown in Sect. 2.2.

A8.2 Derivation of (8.18, 19)

If we write (8.5) as $L(V) = 0$ and define its Green's function $\mathscr{G}$ by

$$L[\mathscr{G}(x, y \,|\, x', y')] = -\delta(x - x')\delta(y - y'), \tag{A8.10}$$

the rf potential V at any point P inside a contour C can be expressed in terms of an integral along C as

$$V_p = \oint_C \left(\mathscr{G} \frac{\partial V}{\partial n} - V \frac{\partial \mathscr{G}}{\partial n} \right) ds, \tag{A8.11}$$

where $\partial/\partial n$ denotes a derivative in the direction normal to the contour as in (A3.4). In the present case, we first rewrite the above equation as

$$V_p = \oint_C \left[\mathscr{G} \left(\frac{\partial V}{\partial n} + \text{j} \frac{\kappa}{\mu} \frac{\partial V}{\partial s} \right) - \text{j} \frac{\kappa}{\mu} \frac{\partial V}{\partial s} \mathscr{G} - V \frac{\partial \mathscr{G}}{\partial n} \right] ds. \tag{A8.12}$$

On the other hand, from the reciprocity relation,

$$\oint_C \mathscr{G}(\partial V/\partial s)\, ds = -\oint_C V(\partial \mathscr{G}/\partial s)\, ds \tag{A8.13}$$

holds. Substituting (8.8) in the text and (A8.13) into (A8.12), we obtain

$$V_p = \oint_C \left[\mathscr{G}(-\text{j}\omega\mu_{\text{eff}}di_n) + V \left(\text{j} \frac{\kappa}{\mu} \frac{\partial \mathscr{G}}{\partial s} - \frac{\partial \mathscr{G}}{\partial n} \right) \right] ds. \tag{A8.14}$$

Equation (8.18) can be derived from the above equation in the same manner as (3.1) was derived from (A3.14 − 16).

Equation (8.19) can be derived similarly, but using a different Green's function, mentioned in the last three lines of Sect. 8.2.3.

A9.1 Derivation of (9.27)

We first show that when k is very large, i.e., when the space under consideration is much greater than the wavelength, we may obtain a differential equation only for ϕ and independent of e and h. From (9.25a) in the text, we obtain

$$\nabla \times E = (\nabla \times e - \mathrm{j}k_0 \nabla \phi \times e)\mathrm{e}^{-\mathrm{j}k_0\phi}. \tag{A9.1}$$

Similarly, for the magnetic field, we obtain from (9.25b)

$$\nabla \times H = (\nabla \times h - \mathrm{j}k_0 \nabla \phi \times h)\mathrm{e}^{-\mathrm{j}k_0\phi}. \tag{A9.2}$$

On the other hand, substituting (9.25a, b) into Maxwell's equations leads to

$$\nabla \times E = -\mathrm{j}k_0 c\mu H = -\mathrm{j}k_0 c\mu h\,\mathrm{e}^{-\mathrm{j}k_0\phi}, \tag{A9.3}$$

$$\nabla \times H = \mathrm{j}k_0 c\varepsilon E = \mathrm{j}k_0 c\varepsilon e\,\mathrm{e}^{-\mathrm{j}k_0\phi}, \tag{A9.4}$$

where c is the velocity of light. Combining (A9.1) with (A9.3), and (A9.2) with (A9.4), we obtain

$$\nabla \phi \times e - c\mu h = (1/\mathrm{j}k_0)\nabla \times e, \tag{A9.5}$$

$$\nabla \phi \times h + c\varepsilon e = (1/\mathrm{j}k_0)\nabla \times h. \tag{A9.6}$$

We now assume that the space under consideration and the lengths associated with the behavior of the ray (e.g., the beam diameter and radius of curvature of the beam axis) are all much greater than the wavelength. In such a case we may consider that the wavelength is extremely short, or the frequency is extremely high, or, in other words, $k_0 \,(= \omega/c)$ tends to infinity. Hence, the right-hand sides of (A9.5, 6) are both zero, so that

$$\nabla \phi \times e - c\mu h = 0, \tag{A9.7}$$

$$\nabla \phi \times h + c\varepsilon e = 0. \tag{A9.8}$$

These equations give relations between the directions of $\nabla \phi$, h, and e; evidently, each of them is normal to the other two. Hence,

$$e \cdot \nabla \phi = 0, \tag{A9.9}$$

$$h \cdot \nabla \phi = 0. \tag{A9.10}$$

We now go back to (9.26). By using (A9.7) and a vector identity, we can rewrite (9.26) as

$$S = (1/2c\mu)\,\mathrm{Re}\{(e \cdot e^*)\nabla \phi - (e \cdot \nabla \phi)e^*\}. \tag{A9.11}$$

However, since the second term within brackets is zero [see (A9.9)] we obtain (9.27).

A9.2 Derivation of (9.32, 33)

From (9.28),

$$|\nabla\phi|^2 = n^2. \tag{A9.12}$$

Taking gradients of both sides, we obtain

$$2\nabla\phi\nabla^2\phi = 2n\nabla n. \tag{A9.13}$$

This equation and (9.29) lead to

$$S\nabla^2\phi = \nabla n. \tag{A9.14}$$

On the other hand, generally

$$\frac{d}{ds} = \sum_{i=1}^{3} \frac{dx_i}{ds} \frac{\partial}{\partial x_i} = \frac{dr}{ds} \cdot \nabla \quad (x_i = x, y, z). \tag{A9.15}$$

Using (9.29, 30) and (A9.14, 15), we obtain

$$\frac{d}{ds}[ns] = \frac{d}{ds}\nabla\phi = \frac{dr}{ds}\nabla^2\phi = s\nabla^2\phi = \nabla n, \tag{A9.16}$$

which is (9.32). Equation (9.33) can be derived immediately from (9.32).

References

Chapter 1

1.1 T. Okoshi, T. Miyoshi: *Planar Circuits* (Ohm, Tokyo 1975) (in Japanese)
1.2 E. W. Mehal, R. W. Wacker: GaAs Integrated Microwave Circuits. IEEE Trans. ED-**15**, 513 – 517 (1968)
1.3 B. H. Sollen, J. Brown: A Centimeter-Wave Parallel Plate Spectrometer. Proc. IEE. **103**, Part B, 419 – 428 (1956)
1.4 T. Okoshi, S. Hashiguchi, M. Matsumoto, M. Kotani: A Novel Method for Measuring Microwave Oscillator Noise. Trans. IECE Jpn. **53-B**, 459 – 464 (1970) (in Japanese)
1.5 S. Mao, S. Jones, G. D. Vendelin: Millimeter-Wave Integrated Circuits. IEEE Trans. MTT-**16**, 455 – 461 (1968)
1.6 H. Bosma: On Stripline Y-Circulation at UHF. IEEE Trans. MTT-**12**, 61 – 72 (1964)
1.7 S. A. Schelkunoff: *Electromagnetic Waves* (Van Nostrand, New York 1943)
1.8 S. Ridella: Analysis of Three-Layer Distributed Structures with n-Terminal Pairs. Proc. Int. Symp. on Network Theory, Belgrade, Yugoslavia, Sept. 1968, pp. 687 – 707
1.9 G. Biorci, S. Ridella: On the Theory of Distributed Three-Layer n-Port Networks. Alta Frequenza **38**, 615 (1969)
1.10 B. Bianco, P. P. Civalleri: Basic Theory of Three-Layer n-Port. Alta Frequenza **38**, 623 (1969)
1.11 P. P. Civalleri, S. Ridella: Impedance and Admittance Matrices of Distributed Three-Layer N-Ports. IEEE Trans. CT-**17**, 392 – 398 (1971)
1.12 B. Bianco, S. Ridella: Nonconventional Transmission Zeros in Distributed Rectangular Structures. IEEE Trans. MTT-**20**, 297 – 303 (1972)
1.13 T. Okoshi: Microwave Planar Circuits, Report of Tech. Group., IECE Japan, No. MW 68 – 69, Feb. 17, 1969 (in Japanese)
1.14 T. Okoshi: Microwave planar circuits, Record of Joint National Concention of Four EE Institutions, Paper No. 1468, March 1969 (in Japanese)
1.15 T. Okoshi: Planar Circuits. J. IECE Jpn. **52**, 1430 – 1433 (1969) (in Japanese)
1.16 T. Okoshi, M. Migitaka, N. Miyazaki: Gunn Oscillator Using Planar Circuit Resonator, Record of National Convention, IECE Japan, Paper No. 708, Nov. 1969 (in Japanese)
1.17 T. Okoshi, T. Miyoshi: The Planar Circuit – A Novel Approach to Microwave Circuitry, Proc. IEEE Int. Conf. on Circuit and System Theory, Kyoto, Japan, Sept. 9 – 11, 1970
1.18 T. Okoshi, T. Miyoshi: The Planar Circuit – An Approach to Microwave IC, European Microwave Conf., Stockholm, Sweden, Aug. 23 – 27, 1971
1.19 T. Okoshi, T. Miyoshi: The Planar Circuit – An Approach to Microwave Integrated Circuitry. IEEE Trans. MTT-**20**, 245 – 252 (1972)
1.20 P. K. Tien: Light Waves in Thin Films and Integrated Optics. Appl. Opt. **10**, 2395 – 2413 (1971)

Chapter 2

2.1 T. Okoshi, T. Miyoshi: *Planar Circuits* (Ohm, Tokyo 1975) (in Japanese)
2.2 K. Kurokawa: *An Introduction to the Theory of Microwave Circuits* (Academic, New York 1969)
2.3 P. M. Morse, H. Feshbach: *Methods of Theoretical Physics*, Part 1 (McGraw-Hill, New York 1953), p. 834
2.4 J. Helszajn, D. S. James: Planar Triangular Resonators with Magnetic Walls. IEEE Trans. MTT-**26**, 95 – 100 (1978)

2.5 R. Chadha, K. C. Gupta: Green's Functions for Triangular Segments in Planar Microwave Circuits. IEEE Trans. MTT-**28**, 1139 – 1143 (1980)

2.6 S. A. Schelkunoff: *Electromagnetic Waves* (Van Nostrand, New York 1943)

2.7 R. Chadha, K. C. Gupta: Green's Functions for Circular Sectors, Annular Rings and Annular Sectors in Planar Microwave Circuits. IEEE Trans. MTT-**29**, 68 – 70 (1981)

2.8 N. Marcuvitz: *Waveguide Handbook* (McGraw-Hill, New York 1951), p. 60

2.9 S. B. Cohn: Problems in Strip Transmission Lines. IRE Trans. MTT-**3**, 119 (1955)

2.10 G. L. Matthaei, L. Young, E. M. T. Jones: *Microwave Filters, Impedance-Matching Networks and Coupling Structures* (McGraw-Hill, New York 1964), p. 440

2.11 T. Okoshi, T. Miyoshi: The Planar Circuit – A Novel Approach to Microwave Circuitry, Proc. IEEE Int. Conf. on Circuit and System Theory, Kyoto, Japan, Sept. 9 – 11, 1970

2.12 T. Okoshi, T. Miyoshi: Analysis of Planar Circuit, Ann. Rep. of Eng. Res. Int., Faculty of Eng., Univ. of Tokyo, Vol. 30, pp. 153 – 168, 1971 (in English)

2.13 T. Okoshi, T. Takeuchi, J.-P. Hsu: Planar 3-dB Hybrid Circuit. Trans. IECE Jpn. **58B**, 408 – 415 (1975)

2.14 T. Okoshi, T. Takeuchi, J.-P. Hsu: Planar 3-dB Hybrid Circuit. Electron. Commun. Jpn. **58**, 80 – 90 (1975) (English transl. of [2.13])

2.15 G. D'Inzeo, F. Giannini, C. M. Sodi, R. Sorrentino: Method of Analysis and Filtering Properties of Microwave Planar Networks. IEEE Trans. MTT-**26**, 462 – 471 (1978)

2.16 R. R. Bonetti, P. Tissi: Analysis of Planar Disk Networks. IEEE Trans. MTT-**26**, 471 – 477 (1978)

2.17 G. D'Inzeo, F. Giannini, R. Sorrentino: Wide-Band Equivalent Circuits of Microwave Planar Networks. IEEE Trans. MTT-**28**, 1107 – 1113 (1980)

Chapter 3

3.1 T. Okoshi, T. Miyoshi: The Planar Circuit – An Approach to Microwave Integrated Circuitry. IEEE Trans. MTT-**20**, 245 – 252 (1972)

3.2 J. A. Stratton: *Electromagnetic Theory* (McGraw-Hill, New York 1941) p. 460

3.3 S. A. Schelkunoff: *Electromagnetic Waves* (Van Nostrand, New York 1943)

3.4 R. F. Narrington: *Field Computation by Moment Methods* (MacMillan, New York 1968)

3.5 M. J. Beaubien, A. Wexler: An Accurate Finite-Difference Method for High-Order Waveguide Modes. IEEE Trans. MTT-**16**, 1007 – 1017 (1968)

3.6 P. L. Arlett, A. K. Bahrani, O. C. Zinkiewicz: Application of Finite Elements to the Solution of Helmholtz's Equation. Proc. IEE **115**, 1762 – 1766 (1968)

3.7 P. Silvester: A General High-Order Finite-Element Waveguide Analysis Program. IEEE Trans. MTT-**17**, 204 – 210 (1969)

3.8 R. H. T. Bates: The Theory of the Point-Matching Method for Perfectly Conducting Waveguides and Transmission Lines. IEEE Trans. MTT-**17**, 294 – 301 (1969)

3.9 R. M. Bulley, J. B. Davis: Computation of Approximate Polynomial Solutions to TE Modes in an Arbitrarily Shaped Waveguide. IEEE Trans. MTT-**17**, 440 – 446 (1969)

3.10 R. M. Bulley: Analysis of the Arbitrarily Shaped Waveguide by Polynomial Approximation. IEEE Trans. MTT-**18**, 1022 – 1028 (1970)

3.11 J. H. Richmond: Digital Computer Solutions of the Rigorous Equations for Scattering Problems. Proc. IEEE **53**, 796 – 804 (1965)

3.12 P. C. Waterman: Matrix Formulation of Scattering. Proc. IEEE **53**, 805 – 812 (1965)

3.13 M. G. Andreasen: Scattering from Cylinders with Arbitrary Surface Impedance. Proc. IEEE **53**, 812 – 817 (1965)

3.14 P. Silvester: Finite-Element Analysis of Planar Microwave Networks. IEEE Trans. MTT-**21**, 104 – 108 (1973)

3.15 J.-P. Hsu, T. Anada, H. Makino, O. Kondo: Computer Analysis of Normal Modes in Planar Circuits, Report of Tech. Group, IECE Japan, No. MW73-101, Nov. 30, 1973 (in Japanese)

3.16 J.-P. Hsu, T. Anada: Planar Circuit Equation and Its Practical Application to Planar-Type Transmission-Line Circuit, 1983 IEEE MTT-S Symposium, June 1983, Technical Digest pp. 574 – 576, Paper No. V-2

3.17 T. Miyoshi, Y. Sakakibara: Eigenfunction Analysis of Planar Circuits and Its Application, Report of Tech. Group, IECE Japan, No. MW74-8, May 24, 1974 (in Japanese)

3.18 N. Marcuvitz: *Waveguide Handbook* (McGraw-Hill, New York 1951), p. 345

3.19 Y. Ayasli: Analysis of Wide-Band Stripline Circulators by Integral Equation Technique. IEEE Trans. MTT-**21**, 200 – 209 (1980)

3.20 S. Kitazawa: Choice of the Green's Function in the Contour-Integral Analysis of Planar Circuits. An internal report in Dept. of Electronic Engineering, Univ. of Tokyo, March 1982 (in Japanese)

Chapter 4

4.1 T. Okoshi, S. Kitazawa: Computer Analysis of Cavity-Type Planar Circuit. IEEE Trans. MTT-**23**, 299 – 306 (1975)

4.2 T. Okoshi, T. Miyoshi: The Planar Circuit – An Approach to Microwave Integrated Circuitry. IEEE Trans. MTT-**20**, 245 – 252 (1972)

4.3 J.-P. Hsu, T. Anada: Planar Circuit Equation and Its Practical Application to Planar-Type Transmission-Line Circuit, 1983 IEEE MTT-S International Microwave Symposium, Paper No.V-2, Tech. Digest pp. 574 – 576

4.4 M. Koshiba, M. Sato, M. Suzuki: Application of Finite-Element Method to H-Plane Waveguide Discontinuities. Electron. Lett. **18**, 364 – 365 (1982)

4.5 M. Koshiba, M. Sato, M. Suzuki: Finite-Element Analysis of Arbitrarily Shaped H-Plane Waveguide Discontinuities. Trans. IECE Jpn. **E66**, 75 – 80 (1983)

4.6 S. Ramo, J. R. Whinnery: *Fields and Waves in Modern Radio* (John Wiley, New York 1953)

4.7 N. Marcuvitz: *Waveguide Handbook* (McGraw-Hill, New York 1951)

4.8 T. Okoshi, S. Kitazawa: Computer Analysis of Short Boundary Planar Circuits (II), Paper of Tech. Group, IECE Japan, No. MW75-75, Oct. 23, 1975 (in Japanese)

4.9 R. E. Collin: *Field Theory of Guided Waves* (McGraw-Hill, New York 1960)

Chapter 5

5.1 T. Okoshi, T. Takeuchi: Analysis of Planar Circuits by Segmentation Method. Trans. IECE Jpn. **58-B**, 400 – 407 (1975) (in Japanese)

5.2 T. Okoshi, T. Takeuchi: Analysis of Planar Circuits by Segmentation Method. Electron. Commun. Jpn. **58**, 71 – 79 (1975) (English transl. of [5.1])

5.3 T. Okoshi, Y. Uehara, T. Takeuchi: The Segmentation Method – An Approach to the Analysis of Microwave Planar Circuits. IEEE Trans. MTT-**24**, 662 – 668 (1976)

5.4 R. Chadha, K. C. Gupta: Segmentation Method Using Impedance Matrices for Analysis of Planar Microwave Circuits. IEEE Trans. MTT-**29**, 71 – 74 (1981)

5.5 V. A. Monaco, P. Tiberio: Computer-Aided Analysis of Microwave Circuits. IEEE Trans. MTT-**22**, 249 – 263 (1974)

5.6 H. Howe, Jr.: *Stripline Circuit Design* (Artech, Dedham, MA 1974), Chap. 3, p. 95

5.7 V. A. Monaco, P. Tiberio: Automatic Matrix Computation of Microwave Circuits. Alta Freq. **39**, 59 – 64 (1970)

5.8 P. C. Sharma, K. C. Gupta: Desegmentation Method for Analysis of Two-Dimensional Microwave Circuits. IEEE Trans. MTT-**29**, 1094 – 1098 (1981)

Chapter 6

6.1 T. Okoshi, T. Miyoshi: The Planar Circuit – An Approach to Microwave Integrated Circuitry. IEEE Trans. MTT-**20**, 245 – 252 (1972)

6.2 K. Grüner: Method for Synthesizing Nonuniform Waveguides. IEEE Trans. MTT-**22**, 317 – 322 (1974)

6.3 F. Kato, M. Saito, T. Okoshi: Computer-Aided Synthesis of Planar Circuits. IEEE Trans. MTT-**25**, 814 – 819 (1977)

6.4 T. Okoshi, Y. Uehara, T. Takeuchi: Segmentation Method – An Approach to the Analysis of Microwave Planar Circuit. IEEE Trans. MTT-**24**, 662 – 668 (1976)

6.5 G. L. Matthaei, L. Young, E. M. Jones: *Microwave Filters, Impedance-Matching Networks and Coupling Structures* (McGraw-Hill, New York 1964), Chap. 13

6.6 N. Marcuvitz: *Waveguide Handbook* (McGraw-Hill, New York 1951), p. 160

Chapter 7

7.1   G. L. Matthaei, L. Young, E. M. Jones: *Microwave Filters, Impedance-Matching Networks and Coupling Structures* (McGraw-Hill, New York 1964), Chap. 13

7.2   T. Okoshi, T. Imai, K. Ito: Computer-Oriented Synthesis of Optimum Circuit Pattern of 3-dB Hybrid Ring by the Planar Circuit Approach. IEEE Trans. MTT-**29**, 194 – 202 (1981)

7.3   E. Polak: *Computational Methods in Optimization*, in Mathematics in Science and Engineering Series, Vol. 77 (Academic, New York 1971)

7.4   G. P. Riblet: An Eigenadmittance Condition Applicable to Symmetrical Four-Port Circulators and Hybrids. IEEE Trans. MTT-**26**, 275 – 279 (1978)

7.5   N. Marcuvitz: *Waveguide Handbook* (McGraw-Hill, New York 1951), Sect. 3.5

7.6   R. F. Collin: *Field Theory of Guided Waves* (McGraw-Hill, New York 1960), Sect. 4.3

7.7   G. P. Riblet: A Directional Coupler with Very Flat Coupling. IEEE Trans. MTT-**26**, 70 – 74 (1978)

7.8   T. Okoshi, T. Miyoshi: The Planar Circuit – An Approach to Microwave Integrated Circuitry. IEEE Trans. MTT-**20**, 245 – 252 (1972)

7.9   K. Grüner: Method of Synthesizing Nonuniform Waveguides. IEEE Trans. MTT-**22**, 317 – 322 (1974)

7.10 K. Kato, M. Saito, T. Okoshi: Computer-Aided Synthesis of Planar Circuits. IEEE Trans. MTT-**25**, 814 – 819 (1974)

7.11 T. Okoshi, Y. Uehara, T. Takeuchi: The Segmentation Method – An Approach to the Analysis of Microwave Planar Circuits. IEEE Trans. MTT-**24**, 662 – 668 (1976)

Chapter 8

8.1   T. Miyoshi, S. Yamaguchi, S. Goto: Ferrite Planar Circuits in Microwave Integrated Circuits. IEEE Trans. MTT-**25**, 593 – 600 (1977)

8.2   T. Miyoshi, S. Miyauchi: The Design of Planar Circulators for Wide-Band Operation. IEEE Trans. MTT-**28**, 210 – 214 (1980)

8.3   H. Bosma: On Stripline Y-Circulation at UHF. IEEE Trans. MTT-**12**, 61 – 72 (1964)

8.4   M. E. Hines: Reciprocal and Nonreciprocal Modes of Propagation in Ferrite Stripline and Microstrip Devices. IEEE Trans. MTT-**19**, 442 – 451 (1971)

8.5   P. de Santis, F. Pucci: The Edge-Guided Wave Circulator. IEEE Trans. MTT-**23**, 516 – 519 (1975)

8.6   H. A. Atwater: *Introduction to Microwave Theory* (McGraw-Hill, New York 1962)

8.7   C. E. Fay, R. L. Comstock: Operation of the Ferrite Junction Circulator. IEEE Trans. MTT-**13**, 15 – 27 (1965)

8.8   Y. S. Wu, F. J. Rosenbaum: Wideband Operation of Microstrip Circulators. IEEE Trans. MTT-**22**, 849 – 856 (1974)

8.9   S. Ayter, Y. Ayasli: The Frequency Behavior of Stripline Circulator Junctions. IEEE Trans. MTT-**26**, 197 – 202 (1978)

8.10 G. P. Riblet: The Measurement of the Equivalent Admittance of 3-Port Circulators via an Automated Measurement System. IEEE Trans. MTT-**25**, 401 – 405 (1977)

8.11 J. Helszajn, D. S. James, W. T. Nisbet: Circulators Using Planar Triangular Resonator. IEEE Trans. MTT-**27**, 188 – 193 (1979)

Chapter 9

9.1   T. Okoshi: *Optical Fibers* (Academic, New York 1982), Chap. 5

9.2   D. Marcuse: *Theory of Dielectric Optical Waveguides* (Academic, New York 1974)

9.3   H. K. V. Lotsch: Physical-Optics Theory of Planar Dielectric Waveguides. Optik **27**, 239 – 254 (1968)

9.4   Y. Suematsu, K. Iga: *Introduction to Optical Fiber Communications* (John Wiley, New York 1982)

9.5   J. Arnaud: *Beam and Fiber Optics* (Academic, New York 1976)

9.6   R. Ulrich, R. Zengerle: Optical Bloch Waves in Periodic Planar Waveguides, Technical Digest, Topical Meeting on Integrated Optics and Guided-Wave Optics, Paper No. TuBl, Jan. 28 – 30, 1980, at Incline Village, Nevada

9.7   A. A. Oliner (ed.): *Acoustic Surface Waves*, Topics Appl. Phys., Vol. 24 (Springer, Berlin, Heidelberg 1978)

9.8   S. Tanaka: Planar Optical Components. Final Report of Special Project Research on Optical Guided-Wave Electronics, Ministry of Education, Japanese Government, pp. 448 – 461, March 1981 (in Japanese)

9.9   S. Tanaka: Optical Waveguide Lenses – A Review. Oyo Butsuri (Applied Optics) **48**, 241 – 248 (1979) (in Japanese)

9.10   R. K. Luneburg: *Mathematical Theory of Optics* (University of California Press, Berkeley, CA 1966), p. 182

9.11   M. Young: *Optics and Lasers*, 2nd. ed., Springer Ser. Opt. Sci., Vol. 5 (Springer, Berlin, Heidelberg 1984)

Chapter 10

10.1   P. K. Tien: Light Waves in Thin Films and Integrated Optics. Appl. Opt. **10**, 2395 – 2413 (1971)

10.2   R. Baets, P. E. Lagasse: Calculation of Radiation Loss in Integrated-Optic Tapers and Y-Junctions. Appl. Opt. **21**, 1972 – 1978 (1982)

10.3   H. K. V. Lotsch: Beam Displacement at Total Reflection: The Goos-Hänchen Effect. Optik **32**, 116 – 137, 189 – 204, 299 – 319, 553 – 569 (1970/71)

10.4   D. Marcuse: *Theory of Dielectric Optical Waveguides* (Academic, New York 1974)

10.5   T. Okoshi: *Optical Fibers* (Academic, New York 1982), Chap. 3

10.6   E. A. J. Marcatili: Dielectric Rectangular Waveguide and Directional Coupler for Integrated Optics. Bell Syst. Tech. J. **48**, 2071 – 2102 (1969)

10.7   C. Yeh, S. B. Dong, W. Oliver: Arbitrarily Shaped Inhomogeneous Optical Fiber or Integrated Optical Waveguides. J. Appl. Phys. **46**, 2125 – 2129 (1975)

10.8   C. Yeh, K. Ha, S. B. Dong, W. P. Brown: Single-Mode Optical Waveguides. Appl. Opt. **18**, 1490 – 1504 (1979)

10.9   M. Ikeuchi, H. Sawami, H. Niki: Analysis of Open-Type Dielectric Waveguides by the Finite-Element Iterative Method. IEEE Trans. MTT-**29**, 234 – 239 (1981)

10.10 N. Mabaya, P. E. Lagasse, P. Vandenbulcke: Finite Element Analysis of Optical Waveguides. IEEE Trans. MTT-**29**, 600 – 605 (1981)

10.11 K. Yasuura, K. Shimohara, T. Miyamoto: Numerical Analysis of a Thin-Film Waveguide by Mode-Matching Method. J. Opt. Soc. Am. **70**, 183 – 191 (1980)

10.12 M. D. Feit, J. A. Fleck, Jr.: Light Propagation in Graded-Index Optical Fibers. Appl. Opt. **17**, 3990 – 3998 (1978)

10.13 M. D. Feit, J. A. Fleck, Jr.: Calculation of Dispersion in Graded-Index Multimode Fibers by a Propagation-Beam Method. Appl. Opt. **18**, 2843 – 2851 (1979)

10.14 H. J. Nussbaumer: *Fast Fourier Transform and Convolution Algorithms*, 2nd. ed., Springer Ser. Inform. Sci., Vol. 2 (Springer, Berlin, Heidelberg 1982)

Subject Index

Applied Physics A

Solids and Surfaces
In Cooperation with the German Physical Society (DPG)
ISSN 0721-7250   Title No. 339

Applied Physics is a monthly journal for the rapid publication of experimental and theoretical investigations of applied research, issued in two parallel series. **Part A** with the subtitle "Solids and Surfaces" mainly covers the condensed state, including surface science and engineering.

Fields and Editors:

Solid-State Physics

Semiconductor Physics: H. J. Queisser, MPI, Stuttgart
Amorphous Semiconductors: M. H. Brodsky, IBM, Yorktown Heights
Magnetism and Superconductivity: M. B. Maple, UCSD, La Jolla
Metals and Alloys, Solid-State Electron Microscopy: S. Amelinckx, Mol
Positron Annihilation: P. Hautojärvi, Espoo
Solid-State Ionics: W. Weppner, MPI, Stuttgart

Surface Sciences

Surface Analysis: H. Ibach, KFA Jülich
Surface Physics: D. Mills, UCI, Irvine
Chemisorption: R. Gomer, U. Chicago

Surface Engineering

Ion Implantation and Sputtering: H. H. Andersen, U. Copenhagen
Device Physics: M. Kikuchi, Sony, Yokohama
Laser Annealing and Processing: R. Osgood, Columbia U. New York
Integrated Optics, Fiber Optics, Acoustic Surface-Waves: R. Ulrich, TU Hamburg

Editor: H. K. V. Lotsch, Heidelberg
Special Features:
– Rapid publication (3–4 months)
– No page charges for concisely written reports
– 50 complimentary offprints

Subscription Information and sample copies are available from your bookseller or directly from Springer-Verlag, Journal Promotion Dept., P.O. Box 105 280, D-6900 Heidelberg, FRG

Springer-Verlag
Berlin
Heidelberg
New York
Tokyo

U. Tietze, C. Schenk

Advanced Electronic Circuits

With the assistance of **E. Schmid**

1978. 570 figures. VIII, 510 pages. ISBN 3-540-08750-8

Contents: Linear and non-linear operational circuitry. – Controlled sources and impedance converters. – Active filters. – Broadband amplifiers. – Power amplifiers. – Power supplies. – Analog switches and comparators. – Signal generators. – Combinatorial logic circuitry. – Sequential logic circuitry. – Microprocessors. – Digital filters. – Data transmission and display. – D/A and A/D converters. – Measuring circuits. – Electronic controllers. – Appendix: Definitions and nomenclature. –

Already in its fourth edition and translated into several languages, this best-seller on electronic circuit design is now available in English. The book deals with the most important applications of analog and digital circuits, from the operational amplifier to the microprocessor. Such a comprehensive range of material has as yet appeared only in books of a review nature. This work differs from these in that the individual topics are treated thoroughly and that it is often more detailed than even the specialized literature. Emphasis is put on the application of the most modern integrated circuits available for the purpose, as only with these components can really dependable and competitive solutions be found. The material is presented in a practical treatment of the subject and is set out in 17 chapters falling into two main groups. The analog section contains chapters in linear and non-linear computing circuits, on active filters, broadband and power amplifiers, on power supplies and signal generators. The digital section deals with combinatorial and sequential logic circuits, with applications of microprocessors, with digital filters and problems of data transmission. Chapters on D/A and A/D converters and on circuits for measurement and control are also included. The book addresses the advanced student with some background in electronics and the practising engineer and scientist.

Springer-Verlag
Berlin
Heidelberg
New York
Tokyo